W9-DIU-129

INTRODUCTION TO INTEGRATED-CIRCUIT LAYOUT

INTRODUCTION TO INTEGRATED-CIRCUIT LAYOUT

BRIAN SPINKS
MOTOROLA, INC.

PRENTICE-HALL, INC., Englewood Cliffs, NJ 07632

Library of Congress Cataloging in Publication Data

Spinks, Brian.
  Introduction to integrated-circuit layout.

  Includes index.
  1. Metal oxide semiconductor field-effect transistors—
Design and construction.   2. Electronic drafting.
I. Motorola, inc.   II. Title.
TK7871.95.S65   1985   621.381′73   84-4738
ISBN 0-13-485418-7
ISBN 0-13-485400-4 (pbk.)

*Editorial/production supervision
  and interior design:* Ellen Denning
Manufacturing buyer: Gordon Osbourne

Printed in the United States of America

10 9 8 7 6 5 4 3 2 1

ISBN   0-13-485418-7
ISBN   0-13-485400-4 {PBK}   01

PRENTICE-HALL INTERNATIONAL INC., *London*
PRENTICE-HALL OF AUSTRALIA PTY. LIMITED, *Sydney*
EDITORA PRENTICE-HALL DO BRASIL, LTDA., *Rio de Janeiro*
PRENTICE-HALL CANADA INC., *Toronto*
PRENTICE-HALL OF INDIA PRIVATE LIMITED, *New Delhi*
PRENTICE-HALL OF JAPAN, INC., *Tokyo*
PRENTICE-HALL OF SOUTHEAST ASIA PTE. LTD., *Singapore*
WHITEHALL BOOKS LIMITED, *Wellington, New Zealand*

CONTENTS

PREFACE

This book originated from material supplied by Motorola, Inc. as part of a junior college drafting course, "Introduction to Integrated Circuit Layout and Design." The course is offered to second-year drafting students who have completed courses in printed-circuit layout and design, electronic schematic drafting, and basic electronics to prepare them with the background required for entry-level positions as IC mask designers. It is to this level that the book is addressed.

The material is normally presented to new employees in the field of integrated-circuit layout when they initially report to work for a semiconductor design company. A portion of the material presented in this book may have been regarded by some manufacturers of semiconductor components as secret. To avoid conflicts with trade secrets, the level of technology presented in this book may lag behind the state of the art by as much as ten years.

The objective of the book is to provide the reader with:

A working vocabulary of the trade

The basic theory necessary for the layout of metal-oxide-semiconductor integrated circuits

A method of translating a logic diagram to a schematic drawing for use in designing an integrated circuit

Techniques for the design of a composite drawing of masks for use in the fabrication of an integrated circuit

The requirements for noncircuit elements of integrated circuits, such as logos, alignment keys, and etch marks

To meet these objectives, the material is organized in such a way as to lead the reader through all aspects of the design of an integrated-circuit mask. A 5-volt N-channel silicon-gate depletion-gate process was chosen for this book. All the exercises and examples deal with this process.

A person who does the drawings for an integrated circuit is known as a designer. A designer proficient in the trade needs only a copy of the topological layout rules to do his or her work. A beginner, unfamiliar with the trade, is totally lost unless some explanation is provided as to the "hows and whys" of the electronics. The material presented covers basic theory in elementary form that pertains to integrated circuits. Although it is necessary to understand this material in order to comprehend how a circuit works, readers should not allow the theory to distract them from the basic objective of being able to convert a logic diagram to a composite drawing. To be a good designer of integrated circuits, much designing practice is required.

The material for this book was adapted solely from sources provided by present and former members of the Microprocessor Products Division Design Staff, Motorola, Inc., in Austin, Texas.

The author wishes to express his thanks to all staff members who assisted or contributed to this undertaking. Direct support for this work came from R. Gary Daniels, head of the Microprocessor Design Group, and Wayne Busfield and Peter Fecik, who are in charge of the layout and design area and who provided much technical assistance. Sheila L. Hancock typed the original manuscript. Mary Cunningham and Joe Murphy read the material and made many valuable suggestions.

BRIAN SPINKS
MOTOROLA, INC.

1

INTRODUCTION

**1.1
WHAT IS AN
INTEGRATED CIRCUIT?**

An *integrated circuit* (IC) is a semiconductor component. These circuits are used with electronic circuitry to perform various linear or digital functions. They are produced using one of the many semiconductor manufacturing processes and contain a variety of transistors. Some of the more popular general classes of semiconductors, based on their method of manufacture, are resistor-transistor logic (RTL), transistor-transistor logic (TTL), integrated injection logic (I^2L), and metal-oxide semiconductor (MOS). Within each of these broad categories there are further classifications based on the use of different formulas. Two of the most popular of the MOS processes at the present time are the *complementary metal-oxide semiconductor* (CMOS) and the *N-channel metal-oxide semiconductor* (NMOS). The major ingredients of both of these MOS processes are aluminum, silicon dioxide, and silicon. The differences are that NMOS uses N-channel insulated-gate field-effect transistors, whereas CMOS uses both P-channel and N-channel insulated-gate field-effect transistors.

This book addresses an N-channel metal-oxide semiconductor, field-effect transistor process that permits the fabrication of very large and complex logic functions onto a single die. A *die* is a rectangular piece of a slice cut from a silicon ingot. The die may also be called a *chip* or a *bar*. A photograph of a die appears in Figure 1.1

FIGURE 1.1 Die highpower photograph of the MC6840 programmable time module. (Photograph courtesy of Motorola, Inc.)

and a drawing of a packaged die or part in Figure 1.2. The terms *mask set*, *circuit*, *die* and *composite* are used interchangeably in this text.

In the early days of integrated circuitry, the IC consisted of a few transistors on a die. The geometries were large and difficult to produce. Processes were simple. Controls and techniques were considered almost the "work" of a magician. Within a short time, techniques were developed and refined such that MOS process equipment became a commonplace off-the-shelf item, and by 1970, 2000 tran-

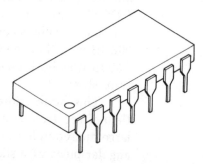

FIGURE 1.2   Integrated circuit in package form.

sistors on a die was typical. Geometries were such that line widths of 10 micrometers (μm; an older term is "micron") were standard. The semiconductor industry dubbed this the age of *medium-scale integration* (MSI). In 1975, the age of *large-scale integration* (LSI) was introduced, where as many as 20,000 transistors were used to make up an IC. Line widths of 6 μm were used throughout the industry. The 1980s brought the introduction of 4-μm line widths and the use of 200,000 transistors to make up an IC. This was the introduction of what is known as *very-large-scale integration* (VLSI). By 1983, 3-μm line widths were appearing in production and 2-μm channel line widths are being built in the laboratory. A new era has arrived. This book is about the techniques used to design a hand-drawn mask set for an integrated circuit.

1.2 MASKS

Each mask contains the patterns necessary for a particular processing step. The process described in Chapter 6 requires eight masks. ICs are produced by creating thin films of silicon dioxide (SiO_2), polycrystalline silicon, metal, and other materials on the surface of a wafer, using masks and a photolithographic process to create a pattern and an etchant to remove portions of the films.

Once the film has been prepared, a layer of photosensitive material is spread evenly over the surface. A mask containing the desired pattern is used to block the exposure of light to selected parts of the photosensitive material. After exposure, this *photoresist* is developed, and the undeveloped photoresist is then removed. The pattern now exists as photoresist on the wafer. If the surface is etched, the photoresist protects that surface beneath it. If the wafer is ion-implanted, the photoresist pattern blocks the ions.

Years ago, masks were created by photographically reducing rubylith artwork. Composite drawings of the masks were drawn 500 times larger (500X) than the actual circuit. These were then photographed and reduced to 300X. The rubylith was cut for each mask layer from the 300X transparent lined photographs. The rubylith was photographed and reduced to 10X on an optically flat plate called a *reticle* (Figure 1.3). The 10X reticle was used to create the mask, which appeared as a multiple of 1X images (Figure 1.4). In the late 1960s and early 1970s special machines called *pattern generators* or *mask-making machines*, built by D. W. Mann, became popular to generate 10X reticles. The reticles were further reduced and stepped to form masks just as before. Today, pattern generators have given way to X-ray and E-beam generators.

Masks are made of optically polished glass plates with an opaque film such as chrome on one surface. The mask pattern is created in the opaque film by a photolithographic process. A photo resist is spread over the opaque film and is exposed using the pattern generator. The photo resist is developed and the pattern is etched. The film that remains has the desired pattern at 10 times the desired size. This mask is called the 10X reticle and is used to create the image for the 1X reticle. The 1X reticle (sometimes the 10X reticle is

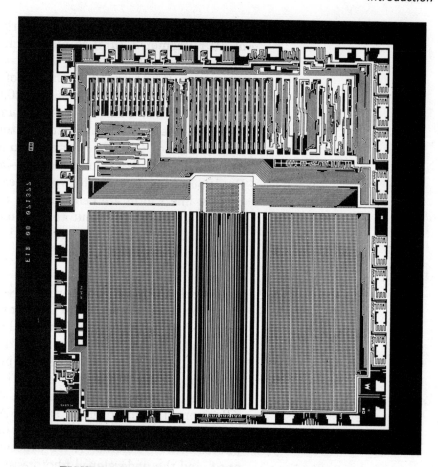

FIGURE 1.3 10X mask or reticle for a single layer.
(Courtesy of Motorola, Inc.)

used instead) is used to create the master mask containing the wafer
pattern. The wafer pattern will consist of as many individual die
patterns as can be "stepped" in the area that will be exposed to the
wafer. From the master, mask working plates referred to as *masks*
are made. Finally, each mask is inspected for quality and proper
dimensions. Certain geometries are compared. The measurement
tolerances are known as *critical dimensions* (CDs). Generally, there
are two classes of masks: dark field or light field, depending on the
wafer photoresist used.

**1.3
PROCESS**

Most integrated circuits are fabricated by growing or depositing
films, and diffusing or ion-implanting dopants into the surface of a
pure silicon wafer. There are many different MOS processes and
there are a variety of flows for each process. A *process flow* consists
of a fixed sequence of processing steps. The process described in this
book is an N-channel silicon-gate metal-oxide-semiconductor process
that uses seven masking steps. Specifics regarding the process flow
will be covered later. This N-channel process allows two different
types of transistors: enhanced and depletion. *Depletion transistors*
are used mainly for loads, whereas *enhanced transistors* are used for

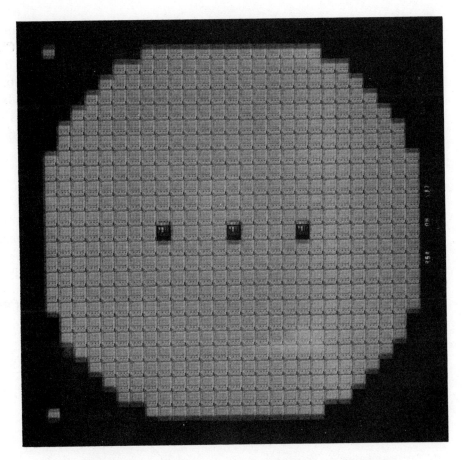

FIGURE 1.4 Single-layer mask. (Courtesy of Motorola, Inc.)

switches, couplers, killers, precharge devices, and sometimes for loads. Because there are N^+, polycrystalline silicon, and metal layers, all of which are conductive, all these layers are used for interconnects. Parasitic capacitance, transistor properties, and resistance are inherent to all conductors.

Processes are usually described in terms of their masking steps. This book describes a process using negative photoresist. In this process, when a wafer is ready for a masking step, one side is covered with a film of viscous, ultraviolet-light-photosensitive material by applying a metered drop of negative photoresist to the wafer while it is spinning. The subsequent film-covered wafer is baked to harden the photoresist. The mask is then aligned with the wafer.

When the pattern image is in the proper position, ultraviolet light is used to expose the unprotected photoresist. It is in this way that the masks are used to pattern the wafers. The wafer is then placed in a developer, which causes the photoresist that was exposed to light to become harder. The unexposed photoresist is then washed away and the surface cleaned and dried, after which the wafer is ready for subsequent processing steps. These steps may include ion implanting and etching. The purpose of this book is to prepare students in the design of such masks.

The masking steps for the N-channel silicon-gate process are defined in the design rules (discussed in Chapter 7). They will include the N⁺ source and drain (01 layer), the depletion implant (02 layer), the buried contact (03 layer), the deposited polycrystalline silicon (05 layer), the preohmic contacts (06 layer), the metal (08 layer), and passivation (09 layer). The process described in this book is a generalized process and does not represent the process of any particular manufacturer.

1.4
LOGIC SYMBOLS:
READING LOGIC

A designer of integrated-circuit masks is expected to be able to reduce logic symbols to schematics on a drawing. In this book the basic MIL-STD 806 logic symbols are used. Special symbols adapted for MOS are used in conjunction with the standard logic symbols. One-third of this book deals specifically with studying the conversion of logic symbols to transistors. Figure 1.5a–c shows some of the symbols that are used in this book, and Figure 1.5d depicts a simple application of a two-input OR gate.

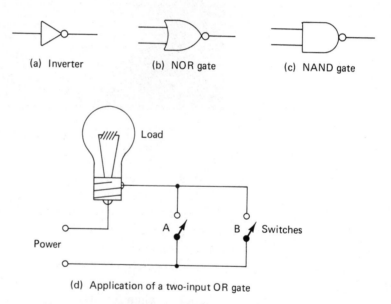

(a) Inverter   (b) NOR gate   (c) NAND gate

(d) Application of a two-input OR gate

FIGURE 1.5 Military Standard 806C common logic symbols and implementation of a two-input or gate.

1.5
TYPICAL DESIGN CYCLE

From the initial concept to the production of prototype parts for a present-day VLSI circuit requires approximately three years of effort (see Figure 1.6). About six months is spent in defining what the integrated circuit, or part, is required to do. A parallel effort, extending for several months, is the development of the part specification. As the specification is being completed, the logic diagram is drawn. It takes six additional months to complete the logic diagram. Sizing the transistors and planning the chip takes about three months. Some critical cells may be designed at this time. Actual circuit layout or mask design takes at least nine months. Mask fabrication and processing requires three months. Once the IC has been fabricated, the parts are characterized and preliminary data sheets are issued. Final circuit "tweaks" and completion of characterization are usually finished at the end of the third year.

1.6 HAND-DRAWN CIRCUITS VERSUS AUTOMATED METHODS

Compared to the hand-drawn circuits described above, it is faster to use gate arrays for creating artwork (or masks). A *gate array* concept is a scheme in which a variety of circuit "building blocks" that may be used to create logic in the form of cells are placed on a die. The cells are connected at the metal masking layer. This technique permits fast production of prototype circuits.

Most of the automated schemes and fast-turnaround techniques tend to create large dies. Even though large dies are expensive to produce, many buyers of ICs want this type of circuit for prototype design or to replace costly components. For IC manufacturers, gate

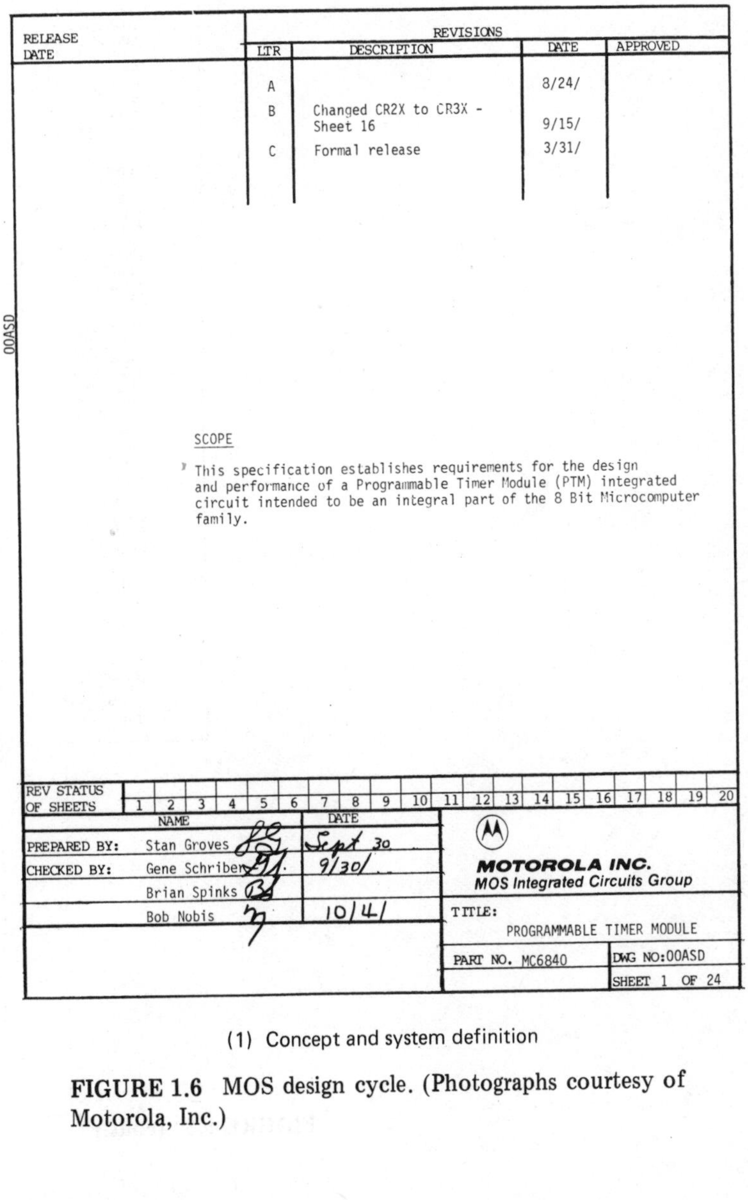

RELEASE DATE

REVISIONS

LTR	DESCRIPTION	DATE	APPROVED
A		8/24/	
B	Changed CR2X to CR3X - Sheet 16	9/15/	
C	Formal release	3/31/	

00ASD

SCOPE

This specification establishes requirements for the design and performance of a Programmable Timer Module (PTM) integrated circuit intended to be an integral part of the 8 Bit Microcomputer family.

REV STATUS OF SHEETS 1 2 3 4 5 6 7 8 9 10 11 12 13 14 15 16 17 18 19 20

	NAME	DATE
PREPARED BY:	Stan Groves	Sept 30
CHECKED BY:	Gene Schriber	9/30/
	Brian Spinks	
	Bob Nobis	10/4/

MOTOROLA INC.
MOS Integrated Circuits Group

TITLE: PROGRAMMABLE TIMER MODULE

PART NO. MC6840 DWG NO: 00ASD

SHEET 1 OF 24

(1) Concept and system definition

FIGURE 1.6 MOS design cycle. (Photographs courtesy of Motorola, Inc.)

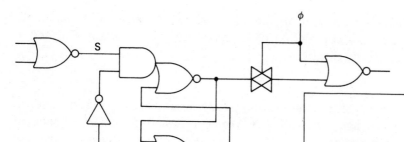

(2) Logic design and simulation

```
                                   S I M U L S Y S
                              MOTOROLA SYSTEM SIMULATOR
                                        AND
                                   TEST GRADER
                                VERSION 3.1 12-7-76

PTM COMPOSITE DECK  VERSION 14 DECEMBER 1976     MARK II
*OPTION:STEPS=1520;MONITOR=A;POPT=1;STATIC.
*PRINT:DAB,G2,O2,C2,G3,O3,C3,  RES,IRQ,RSO,RS1,RS2,RW,INHI,CSO,CS1, PH2,DB7,
C3X,C8D,C8E,C8F,C8G,CI3J,C8EA,C8EC,C8ED,C8EE,C8FA,C8FC,C8FD,C8FE,
DB5,DB4,DB3,DB2,DB1,DBO,G1,O1,C1.
*MACRO(CLGN). A(NOTO)=1. B(ORO)=A,H. C(ORO)=A,2. *DATA:PRNT. D(NORO)=G,F.
E(NOTO)=O. F(NOTO)=A. G(OT1)=P2. H(OT1)=P1. W(ORO)=A,D,3. *DATA:PRNT.
P2(NORO)=A,B. *DATA:PRNT. P1(NORO)=E,F. *DATA:PRNT. *MEND.
*MACRO(CNTR). A(ANDO)=D,Z. B(ANDO)=1,2. D(NOTO)=2. E(ORO)=2,D. G(NOR5)=E,A,B
H(ANDO)=9,Y. I(ANDO)=6,Z. K(ORO)=6,9. M(NOR5)=K,H,I. N(CUPO)=8,M. O(NOTO)=N.
S(MOTO)=O. P(ANDO)=s,7. Q(CUPO)=O,9. R(NORO)=4,5. S(ORO)=4,s. T(ANDO)=M,8.
U(ANDO)=Q,R. Y(NOR5)=W,T,U. W(ORO)=R,8. Z(NORO)=3,G. *MEND.
*MACRO(MSB). A(ANDO)=D,s. B(ANDO)=1,2. D(NOTO)=2. E(ORO)=D,2. s(OR5)=E,A,B.
*MEND.
*MACRO(LSB). B(ORO)=4,5,6. D(NOR5)=O.1,2,3. E(CUPO)=7,D. F(NOTO)=8. G(ANDO)=
H(ANDO)=F,M. J(ORO)=8,F. L(NOR5)=J,G,H. s(ANDO)=9,L. M(NOTO)=L. *MEND.
*MACRO(CRQC). A(ANDO)=D,I. B(ANDO)=1,2. D(NOTO)=2. E(ORO)=2,D. G(NOR5)=E,A,B
s(MOTO)=I. I(NORO)=3,G. *MEND.
*MACRO(CRQB). A(ANDO)=D,G. B(ANDO)=1,2. D(NOTO)=2. E(ORO)=2,D. F(OR5)=E,A,B.
G(ORO)=3,F. s(NOTO)=G. H(NOTO)=s. *MEND.
*MACRO(CRQ
G(ORD)=3,F
*MACRO(IO)
I(NOT1)=6.
*MACRO(CAP
s(NOTO)=E.
*MACRO(LAT
G(CUP1)=1,
*MACRO(DIV
E(NOR5)=B,
*MACRO(SYN
H(ANDO)=2,
*MACRO(DEL
INHI(GFN).
PH2(CLOK).
C(GEN). *D
A(CLOK). *
B(CLOK). *
D(CLOK). *
DAB(ORO)=C
*VECTOR:RE
*DATA: PAR
0100000000
1100000001
1100000000
1100000000
1100000000
1100000000
1100000000
1100000000
1100000000
1100000000
GATE(GEN),
O(GEN). *D
CK(CLGN)=P
CSA(NOTO)=
CSB(NOTO)=
CSC(NOTO)=
CSD(NORO)=
CSE(NORO)=
CSR(NOTO)=
CSW(NOTO)=
C1A(CRQC)=
C1B(CRQA)=
C1C(CRQA)=
C1D(CRQA)=
C1E(CRQA)=
C1F(CRQA)=
C1G(CRQA)=
C1H(CRQA)=
C2A(CRQB)=
C2B(CRQA)=
C2C(CRQA)=
C2D(CRQA)=
```

```
                                                     **** CELLS US

     A    A1A   A1B   A1C   A1D   A1F   A1V   A1W   A2A   A2B
    A3V   A3W    B     C   CE1A  CE1B  CE1C  CE1E  CE1F  CE1G
   CE2F  CE2G  CE2H  CE2I  CE2J  CE2S  CE2T  CE2U  CE3A  CE3B
   CI1A  CI1B  CI1C  CI1D  CI1E  CI1F  CI1G  CI1H  CI1I  CI1J
   CI2H  CI2I  CI2J  CI2K  CI2L  CI2M  CI3A  CI3B  CI3C  CI3D
   CKB   CKC   CKD   CKE   CKF   CKG   CKH   CKP1  CKP2  CKW
   C1A   C1AA  C1AB  C1AD  C1AE  C1AG  C1AI  C1B   C1BA  C1BB
   C1CG  C1D   C1DA  C1DB  C1DD  C1DE  C1DF  C1DG  C1E   C1EA
   C1FF  C1FG  C1G   C1GA  C1GB  C1GD  C1GE  C1GF  C1GG  C1H
   C1KB  C1KD  C1KF  C1KG  C1KH  C1KJ  C1L   C1LA  C1LB  C1LD
   C1PC  C1Q   C1R   C1S   C1T   C2    C2A   C2AA  C2AB  C2AD
   C2BG  C2C   C2CA  C2CB  C2CD  C2CE  C2CF  C2CB  C2D   C2DA
   C2EF  C2FG  C2F   C2FA  C2FB  C2FD  C2FE  C2FF  C2FG  C2G
   C2HE  C2H
   C2LJ  C2M
   C3B   C3A
   C3DG  C3E
   C3GF  C3G
   C3LA  C3L                                              PTM C
   C3W   C3X
   C8FB  C8F          DGOCGOCRIRRRRICCPDDCCCCCCCCCCCCCCDD
   C8J   C8T          A222333ERSSSWNSSHBB3888818888888888B3
   DR2O  DB3           B        SQ012 H01276XDEFG3EEEEFFFF54
   DB6O  DB7                    I            JACDEACDE
   G1CA  G1C
   G1HB  G1H         0              M  P
   G2DA  G2D         9    A  D  G I  M  P     T   YZA CDE G
   G3    G3A        19    A  D  G I  M  PQ   TU   Y A  DE G
   G3F   G3G        29       D  G I  M  PQ   TU   Y A  DE G
   IDH   IF1        39    A  D  G I  M  P     T   ZA CDE G
   IF2A  IF2        49    A  D  GHI  M  P     T   ZA CDE G
   IF3B  IF3        59    A  D  GHI  M  PQ   TU   AB DE G
   LA1L  LA1        69    A  D  GHI  M  PQ   TU   AB DE G
   LBBF  LBR        79    A  D  GHI  M  P    T V  Z B  EF
   LRDF  LBO        89    A  D  GHI  M  P    T V  Z B  EF
   LAFF  LBF        99    A  D  GHI  M  PQ   UV    C EF
   LRHF  LBH       109    A  D  GHI  M  PQ   UV    C EF
   MABD  MBR       119    A  D  GHI  M  P     U   Y A  DE G
   MRFD  MBF       129    A  D  GHI KL      U   Y A  DE G
   OF1I  OF1       139    A  D  GHI KL  PQ   U   Y A  DE G
   OF2Q  OF2       149    A  D  GHI KL  P     U   Y A  DE G
   O1    O2        159    A  D  GHI KL  P     U   Y A  DE G
   RC1   RC2       169    A  D  GHIJKL     U   Y A  DE G
   RSC   RSD       179    A  D  GHIJKL  PQ   U   Y A  DE G
   RSZF  RSZ       189    A  D  GHIJKL  P     U   Y A  DE G
   RS1P  RS1       199    A  D  GHIJKL  P     U   Y A  DE G
   RS3C  RS3       209    A  D  GHI       P RS U   Y A  DE G
   RWB   RWC       219    A  D  GHI       PQRSTU   Y A  DE G
   S7    TE1       229    A  D  GHI       PQRSTU   Y A  DE G
   TE2E  TE2       239    A  D  GHI       P RSTU   Y A  DE G
   T01A  T01       249    A  D  HIJ       P     T   YZA CDE G
   T1AB  T1A       259    A  D  HIJ       PQ    T   YZA CDE G
   T1BA  T1B       269    A  D  HIJ       PQ    T   YZA CDE G
   T1C   T1C       279    A  D  HIJ       P     T   YZA CDE G
   T1CZ  T1D       289    A  D  GHI       P    TU   Y A  DE G
   T1DY  T1D       299    A  D  GHI       PQ   TU   Y A  DE G
   T1EW  T1E       309    A  D  GHI       PQ   TU   Y A  DE G
   T1FU  T1F       319    A  D  GHI       P    TU   Y A  DE G
   T1GT  T1G       329    A  D  GHI  M  P    TU   Y A  DE G
   T1HS  T1H       339    A  D  GHI  M  PQ   TU   Y A  DE G
   T1IR  T1I       349       D  GHI  M  PQ   TU   Y A  DE G
   T1JQ  T1J       359    A  D  GHI  M  P    TU    AB DE G
   T1KP  T1K       369    A  D  GHI  M  P    TU    AB DE G
   T1LO  T1L       379    A  D  GHI  M  PQ   TU    AB DE G
   T1MN  T1M       389       D  GHI  M  P    TU    AB DE G
   T1NM  T1N       399    A  D  GHI  M  P    TU    AB DE G
   T1OK  T1O       409    A  D  HI   M  P     T W  ZA CD F
   T1PI  T1P       419    A  D  HI   M  PQ    T VW Z B  EF
   T2AH  T2A       429       D  HI   M  PQ    X  ZA CDE G
   T2BG  T2R       439    A  D  HI   M  P     V X Z B  EF
   T2CE  T2C       449    A  D  HI   M  P     WX ZA CD F
   T2DD  T2D       459       D  HI   M  PQ    T   ZA CDE G
   T2EB  T2E       469       D  HI   M  PQ    T   ZA CDE G
   T2FA  T2F       479    A  D  HI   M  P    T V  Z B  EF
   T2G   T2G       489    A  D  HI   M  P    T W  ZA CD F
   T2GZ  T2H       499    A  D  HI   M  PQ   T VW Z B   G
```

FIGURE 1.6 (cont.)

(3) Breadboard and logic verification

FIGURE 1.6 (cont.)

arrays do reduce much of the front-end engineering cost, but test development costs still exist.

Another automated technique is the use of predesigned logical building blocks. This technique is called a *programmed placement and routing program.* The blocks contain AND, NAND, NOR, OR, and exclusive-OR gates and buffers in the form of cells. Computer programs are used to place or select the appropriate cell based on fan-out and interconnection of the logical elements. This technique provides a smaller die than that produced by the gate array scheme since the logical elements needed are used. Standard library logical elements are used; therefore, area and routing are not optimum. A mask designer will spend a great deal of time optimizing circuit performance and reducing die area. As a result, hand-drawn circuits are very expensive and tend to produce smaller dies. This process is very good for yield and the cost per die is less, qualities that are necessary for high-volume producers of ICs. Interactive graphic computer terminals are in use today which allow drawing the circuit "on the terminal" or directly into the computer data base. This book addresses how to design and lay out hand-drawn masks or "circuits." Some custom manufacturers provide quick-turnaround, high-cost circuits, such as gate arrays, for prototypes. If the circuit goes into mass production, the circuit is then hand-drawn to increase the yield and lower the manufacturing costs.

Note that the gate array scheme produces the largest die size. Custom-designed hand-drawn circuits tend to be smallest and most performance efficient. The programmed routing placement program produces circuits from precoded drawn cells. These circuits are somewhat inefficient in terms of area, but better than those produced by the gate array scheme.

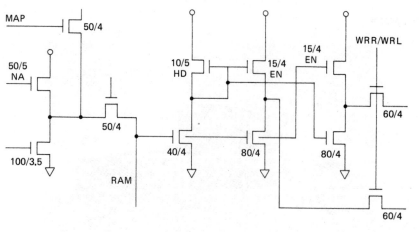

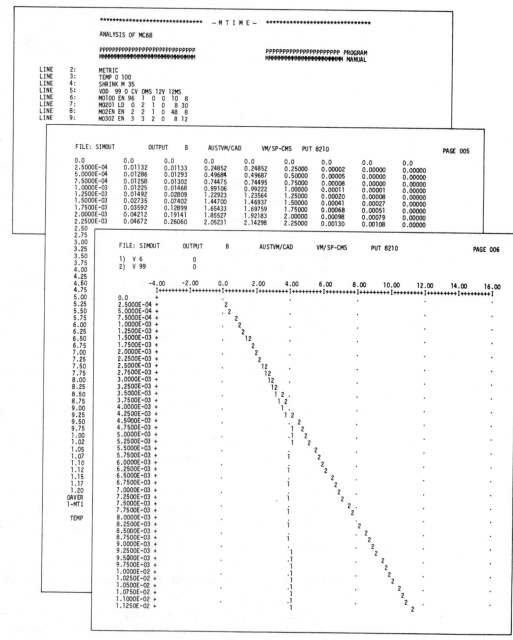

(4) Circuit design and analysis

FIGURE 1.6 (cont.)

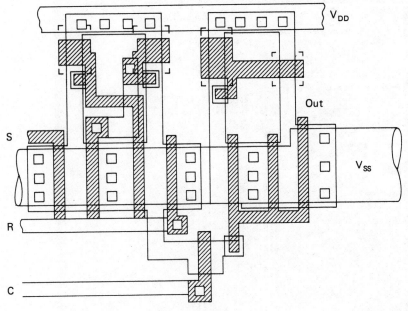

(5) Composite layout

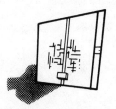

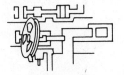

(6) Digitizing, editing, composite
verification

FIGURE 1.6 (cont.)

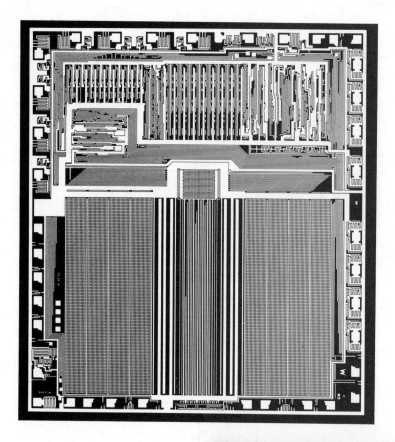

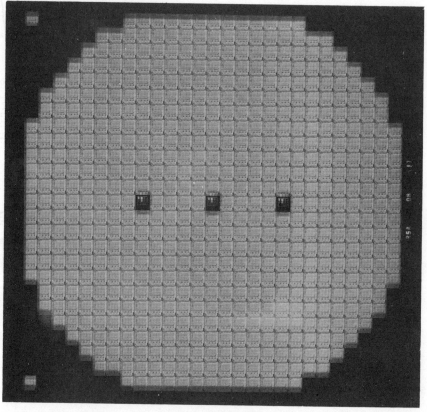

(7) Mask generation

FIGURE 1.6 (cont.)

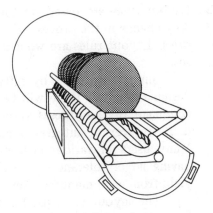

(8) Wafer processing

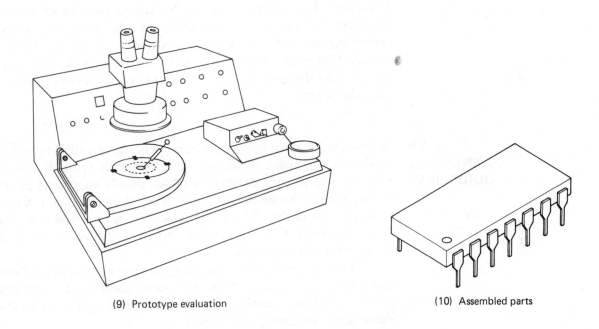

(9) Prototype evaluation   (10) Assembled parts

FIGURE 1.6 (cont.)

**1.7
SCHEDULES**
As we all know, time is money. In the semiconductor business, time can be the difference between success or failure. Schedules are established for designs on the basis of the need to enter the marketplace at the appropriate time. A slip in schedule may mean a loss of market. Design schedules are based on the history of previous designs, market need, the complexity of the circuitry, and the number of designers assigned to the project.

Related to the matter of schedules is the fact that most graphics are done with the aid of a computer. All coordinates and line lengths must be given numbers. The assigning of numbers is called *digitizing*. Since most designs are digitized or "drawn on the tube," speed and accuracy are more important than neatness. Nevertheless, the drawings must be legible.

**1.8
LAYOUT RULES**

Layout rules are based on the characteristics of a specific process. Any change in the process will change the requirements of the *layout rules*. Layout rules are what designers use when they design a mask set.

Similar processes by different manufacturers may have similar design rules, but there will be different requirements. Only persons cognizant of the process know its limitations and weaknesses. Each manufacturer of integrated circuits will have its own set of rules. Many users of various custom manufacturers use generalized rules having larger tolerances, a procedure that tends to produce a high yield from each manufacturer.

The layout rules used in this text are generalized. The process is an N-channel self-aligning silicon-gate process. Design engineers specify the design rules to be used for the circuit design. When designing a circuit, designers are expected to use minimum dimensions wherever it is advantageous either for size or performance. To assist designers in avoiding violations of layout rules, most companies have computers and software which can do logic checks and design rule checks. Computer programs may be used to do the initial layout designs. The designer will take advantage of the layout rules to optimize the design.

**1.9
ENGINEERING
GUIDELINES**

In addition to providing the designer with a logic diagram and a *cell book* which explains the details of certain logic symbols, the design engineer will provide device-size tables, details on critical paths and nodes, pin-out requirements, and other things that are needed to complete the design. The design engineer, who works with the designer to produce a chip plan, is responsible for the sizes of all the transistors and power buses.

The detailed data that the design engineer provides the designer, apart from the logic and the cell book, comprise the *engineering guidelines*. These guidelines should be written and given to the designer at the beginning of a project.

**1.10
PROPERTIES OF
ELECTRONIC DEVICES**

It is assumed that readers have some knowledge of high-school-level physics, including a working knowledge of basic electronics in relation to:

1. Properties of electronics
2. Electric current
3. Electrostatic force
4. Capacitance
5. Ohm's law
6. Resistance
7. Properties of transistors
8. Electric power

9. Electric potential
10. Diodes

Following is a brief summary of these topics.

An *electron*, an elementary particle that is a fundamental constituent of matter, has a negative charge and is located outside the nucleus of an atom.

Electric current is the flow of electrons. A positive flow of current is in the direction opposite to the flow of electrons. Holes that occur in semiconducting material are vacancies in the valance band of the atoms making up the material.

Electrostatic force is the attracting or repelling force caused by the presence or lack of electrons in the proximity of other electrons or lack of electrons. It is expressed by the formula

$$\text{ESF} = \frac{kq_1q_2}{d^2}$$

where k is a constant, q_1 and q_2 are the two charges, and d is the distance between the charges.

Capacitance is the property of being able to collect and store a charge of electricity.

Ohm's law describes the relationship between resistance, voltage, and current.

Resistance is the property of matter to oppose the flow of current.

The *properties of transistors* are such that their resistance to the flow of current through two electrodes may be controlled by one or more additional electrodes.

Electric power is the ability to perform work.

Electric potential is the electromotive force that will cause current to flow.

A *diode* comprises two electrical terminals that have the property of allowing current to flow in only one direction. Diodes may be formed by the junction of P- and N-type semiconductor material. This characteristic allows transistors, interconnect diodes, and resistors to be formed in a silicon substrate or base. All that is required is that these junctions be reversed-biased.

Capacitors and Silicon

Capacitors consist of two plates that must be capable of holding a charge (Figure 1.7). In the ideal capacitor, the plates are conductive; however, the plates may be nonconductive. If they are nonconductive, the charge will be on the surface. A charge on the plates will create an *electric field*, E. The strength of the field is proportional

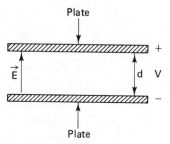

FIGURE 1.7 The elements of a simple capacitor.

to the voltage applied and inversely proportional to the distance between the plates.

Any charges between the plates will be acted upon by the field. (The charges will also contribute to the overall field.) The *force* on the charge is proportional to the field:

$$\text{force (on electron, } q) = -q \cdot \mathbf{E}$$

where q is the charge on an electron.

Transistors

Matter has electrical properties and is classified as either a conductor or a nonconductor. Materials that may be conductive and nonconductive at the same time are called *semiconductors*. Pure silicon is such a material. A transistor is a solid-state electronic device composed of semiconductor material, such as silicon, that controls current flow. Transistors have the advantages of being of low power, long-lived, and compact.

The resistance of pure silicon is high, a characteristic of nonconductors. Applying a field across a piece of silicon will cause the electrons to move, a characteristic of conductors. If the silicon has a small concentration of certain impurities (1 part impurity to 10^{16} parts silicon), it becomes very conductive. It is possible to introduce impurities into pure silicon. This may be done by diffusing, a process in which impurities are placed next to the surface of the silicon to allow the molecules of the impurity to work their way into the silicon over time at elevated temperatures; or by ion implanting, a process in which atoms are accelerated at high speed and allowed to penetrate the surface of the silicon; or by a combination of the two. Diodes are made using one of these processes.

Transistors are built using practically pure germanium or silicon. The elements or electrodes that comprise the transistor may be created by introducing impurities into the germanium or silicon. The introduction of the impurities is called *doping*. The impurity is called the *dopant*. The dopant changes the characteristics of the germanium or silicon. Ordinarily, the silicon used is the substrate for building the transistors. It is lightly doped. For our process, we will use a lightly doped P material, designated P^-. Highly doped N material, designated N^+, may be used for diode junctions.

There are two common types of transistors: bipolar and field-effect. *Bipolar transistors* have emitters, collectors, and bases, and utilize the biasing of junctions for their function. *Field-effect transistors* have sources, drains, and gates, and utilize the effect of an electric field for their function. Integrated circuits may be designed using either or both types of transistors. (However, when bipolar transistors have been designed for use in an MOS technology, their efficiency has not been very high.)

MOSFETS

There are two types of field-effect transistor: the *insulated-gate metal-oxide-semiconductor field-effect transistor* (MOSFET) and the *junction field-effect transistor* (JFET). Metal-oxide-semiconductor field-effect transistors are different from junction field-effect transistors in that the control gate is insulated from the channel. The gate is, in effect, a capacitor. Diode junctions may be used as interconnections for transistor logic and power, and source and drains. When N-doped silicon having an excess of electrons comes in contact with P-doped silicon having a lack of electrons, the difference in contact potential creates an electric field. This electric field creates a *depletion region*, a region in which there are no mobile charges.

Elements of MOSFETS. All MOS transistors are composed of sources, drains, gates, and bulk. The drain for N-channel transistors is the more positive element of the two elements: the source and the drain (Figure 1.8). (For P-channel transistors, it will be the most negative element.) A cross-sectional view of an N-channel transistor as it might appear is shown in Figure 1.9.

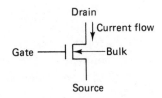

FIGURE 1.8   The elements of the MOS transistor.

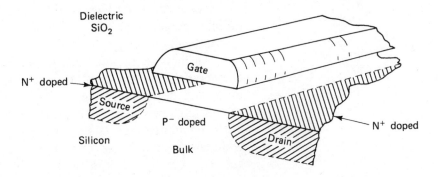

FIGURE 1.9   A three-dimensional cross section of a MOS transistor.

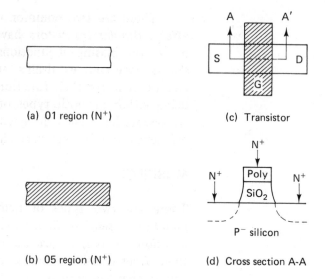

(a) 01 region (N⁺)

(c) Transistor

(b) 05 region (N⁺)

(d) Cross section A-A

FIGURE 1.10 Overlapping the 01 mask layer with 05 mask layer forms a MOS transistor.

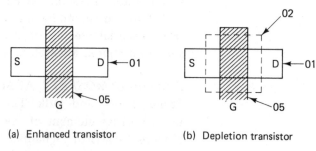

(a) Enhanced transistor

(b) Depletion transistor

FIGURE 1.11 Difference between the enhance transistor and depletion transistor is the 02 layer.

Drawing a transistor. Figure 1.10a and b shows the first two steps in drawing a transistor. The intersection of an 01 and an 05 region forms a transistor (Figure 1.10c). The cross section A-A as it would appear on the die is shown in Figure 1.10d. To differentiate between the enhanced and the depletion transistor, an 02 mask is drawn around the transistor (Figure 1.11).

How an enhanced MOSFET works. As the gate voltage is increased, the electric field (E) increases in the channel region. This causes the mobile charges to gather at the surface of the substrate at the SiO_2-Si interface.

An increase in gate voltage causes the P⁻ material to change its characteristics. Because of the excess electrons, it begins to behave more and more like N⁺ material. As the gate voltage increases, the silicon in the channel takes on the characteristics of intrinsic silicon. With the addition of gate voltage the silicon in the gate behaves like N-doped silicon. When current begins to flow in the channel, the voltage on the gate is known as the threshold voltage, V_{T0}.*

---

*Threshold voltage for discrete MOS transistors is determined by extrapolation. The gate and drain are connected together. Voltage measurements are

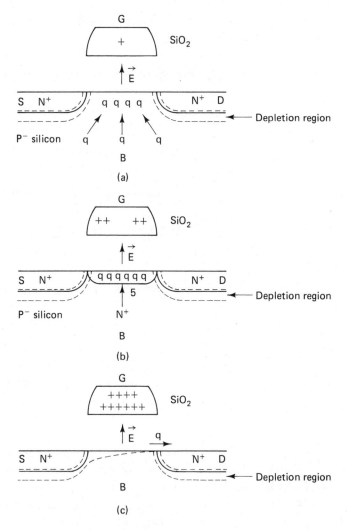

FIGURE 1.12 Changes in the transistor channel resulting from increases in gate voltage.

First, a small amount of charge appears on the gate (Figure 1.12a). As the charge on the gate (G) increases, the voltage with respect to the bulk increases, drawing electrons, 9, to the surface (Figure 1.12b). More charges on the gate increases the voltage further and will cause a complete change in the properties of the bulk silicon near the surface (Figure 1.12c). Additional voltage on the gate causes inversion to occur. At this point, the channel becomes conductive. The gate voltage at which current begins to flow between the drain and source when the source and drain are biased is the threshold voltage, V_{T0}. Voltage on the source and drain also sets up a field. This can cause effects on the gate field. Most V_{TD} measurements are

taken for two drain currents, e.g., 10 mA and 20 mA. These values are plotted. A straight line is drawn through the points to determine the voltage value for zero drain current. That value is known as V_{T0} or threshold voltage. The terms V_P, V_{TD} are used for depletion transistor pinchoff voltage. V_{TE}, V_T are used for enhancement transistor threshold voltages.

made with small V_{DS} to reduce the effect of the source and drain fields.

Points to remember. The masks and their functions are as follows:

01	N^+-doped regions form interconnect sources and drain regions.
05	Polycrystalline silicon regions form interconnects and gates for transistors.
01, 05	Transistors are formed when the 01 and 05 regions intersect.
01	Regions are part of the substrate and form diodes with respect to the bulk region.

Note also the following points.

Gate oxide film exists beneath the polycrystalline silicon regions.

Doping of the N^+ 01 regions and the 05 polycrystalline silicon regions is done after the 05 masking step and is done from the top.

**1.11
TOLERANCES
AND ALIGNMENT**

Drawings, masks, photoresist, and etched product all have tolerances. These are based on history, the capability of the equipment, and the skill of the operators. The mask alignment tolerances, together with the processing tolerances, are incorporated into the topological layout design rules.

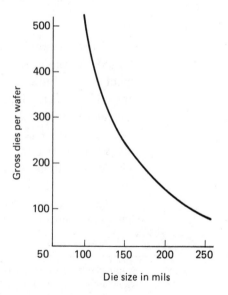

FIGURE 1.13 Relationship of yield to die size: 3-inch wafers.

1.12
YIELD VERSUS DIE SIZE

Smaller die (or chip) size will produce more potential die per wafer. This implies that the yield of good dies per wafer will be higher. Good yield means lower production costs.

Yield tends to vary inversely proportional to the area of the die (Figure 1.13). Note that on the curve of die size versus gross dies per wafer, reducing the die size from 150 mils on a side to 125 mils on a side almost triples the yield.

1.13
CHIP PLANS

The best place to reduce die size is with a good chip plan. A *chip plan* is a plan showing the placement of logic as it is to be organized and located on a mask set. Proper organization and partitioning of the logic on the die will reduce the number of interconnects and the power busing. Chip plans must be taken seriously. Simple things such as routing a power bus can cost or save as much as 20% of the chip area.

1.14
OPTICAL SHRINK

Another technique associated with hand-drawn circuits to improve the gross yield is optical shrink. An *optical shrink* of masks is nothing more than reducing the size of the masks. Optical shrinks work well for ICs that contain a large amount of circuitry. If there is a small amount of circuitry, little is accomplished with an optical shrink. The reason is that the bonding pads must be enlarged prior to an optical shrink so that the finished product has pads that meet the minimum requirements for the wire bonding machinery. Thus the shrunken die has practically the same area as the unshrunken die.

1.15
ACCURACY

Accuracy *cannot* be overstressed. A simple error such as a missing contact can make or break the operation of a circuit. New mask sets cost time and money. Errors annoy engineers and management.

Check and recheck your work. Have someone else verify your designs. Errors are costly. A mask set can cost as much as $1000 per layer, or $8000. Lost time, material costs, and troubleshooting can make a missing contact cost as much as $100,000 and six months' delay in bringing a mask set to production.

1.16
COLOR CODES

To make the implementation of a design easier to recognize for the designer, engineer, digitizer, and checkers, the N^+ regions and the polycrystalline silicon regions are colored. The following convention will be used:

Region	Color
N^+ tied to V_{DD}	Light blue
N^+ tied to V_{SS}	Orange
N^+ tied to signal nodes	Yellow
Polycrystalline silicon	Green
Metal	Red
Cell boundaries	Dark blue

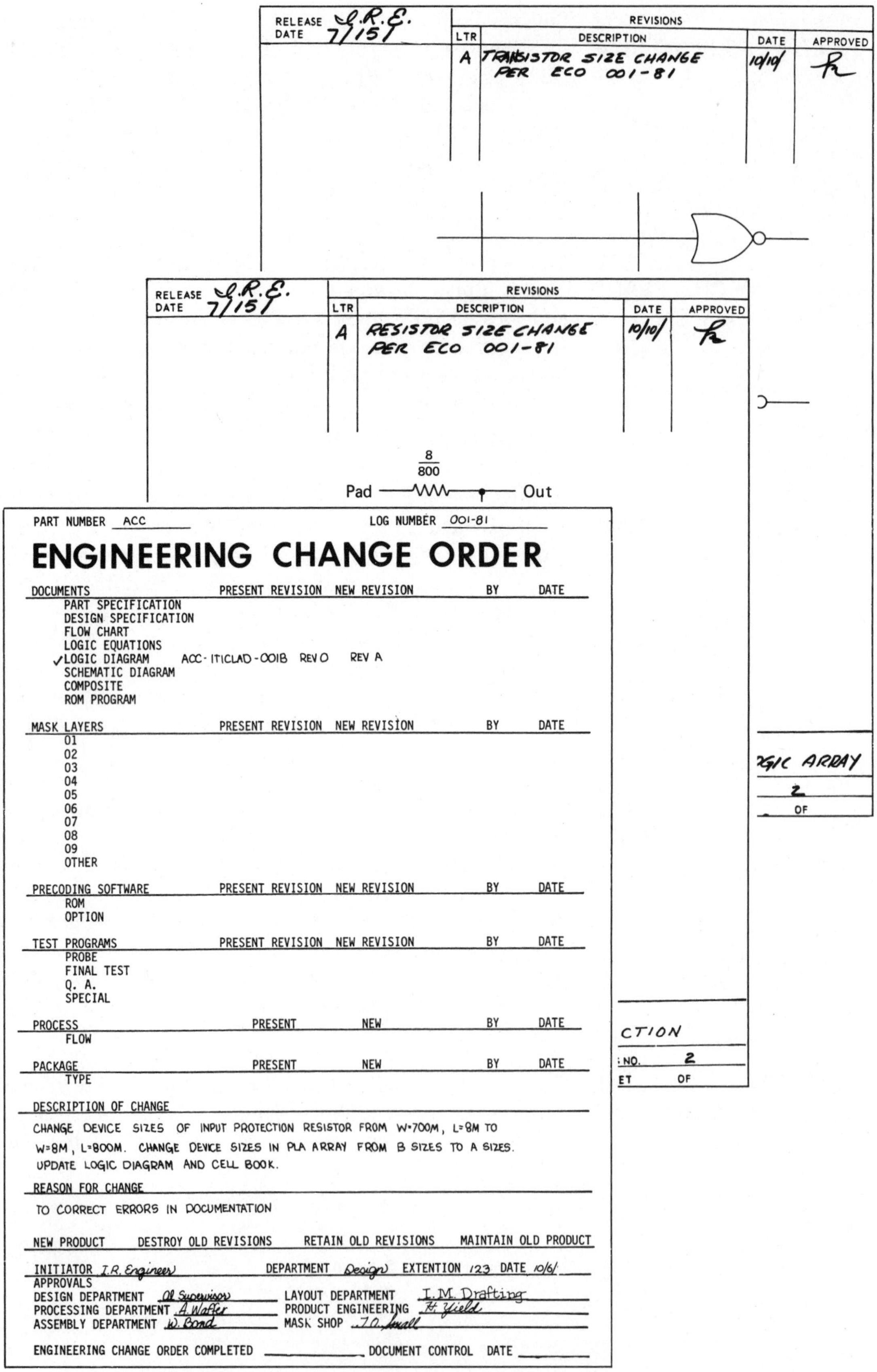

RELEASE	$\mathcal{G.R.E.}$		REVISIONS		
DATE	7/15/	LTR	DESCRIPTION	DATE	APPROVED
		A	*TRANSISTOR SIZE CHANGE PER ECO 001-81*	10/10/	R

RELEASE	$\mathcal{G.R.E.}$		REVISIONS		
DATE	7/15/	LTR	DESCRIPTION	DATE	APPROVED
		A	*RESISTOR SIZE CHANGE PER ECO 001-81*	10/10/	R

$\frac{8}{800}$

Pad ——Λ/Λ/Λ—•—— Out

PART NUMBER ACC   LOG NUMBER 001-81

ENGINEERING CHANGE ORDER

DOCUMENTS	PRESENT REVISION	NEW REVISION	BY	DATE
PART SPECIFICATION				
DESIGN SPECIFICATION				
FLOW CHART				
LOGIC EQUATIONS				
✓LOGIC DIAGRAM	ACC-ITICLAD-001B REV O	REV A		
SCHEMATIC DIAGRAM				
COMPOSITE				
ROM PROGRAM				

MASK LAYERS	PRESENT REVISION	NEW REVISION	BY	DATE
01				
02				
03				
04				
05				
06				
07				
08				
09				
OTHER				

PRECODING SOFTWARE	PRESENT REVISION	NEW REVISION	BY	DATE
ROM				
OPTION				

TEST PROGRAMS	PRESENT REVISION	NEW REVISION	BY	DATE
PROBE				
FINAL TEST				
Q. A.				
SPECIAL				

PROCESS	PRESENT	NEW	BY	DATE
FLOW				

PACKAGE	PRESENT	NEW	BY	DATE
TYPE				

DESCRIPTION OF CHANGE

CHANGE DEVICE SIZES OF INPUT PROTECTION RESISTOR FROM W=700M, L=8M TO
W=8M, L=800M. CHANGE DEVICE SIZES IN PLA ARRAY FROM B SIZES TO A SIZES.
UPDATE LOGIC DIAGRAM AND CELL BOOK.

REASON FOR CHANGE

TO CORRECT ERRORS IN DOCUMENTATION

NEW PRODUCT   DESTROY OLD REVISIONS   RETAIN OLD REVISIONS   MAINTAIN OLD PRODUCT

INITIATOR *I.R. Engineer*   DEPARTMENT *Design* EXTENTION 123 DATE 10/6/
APPROVALS
DESIGN DEPARTMENT *Al Supervisor*   LAYOUT DEPARTMENT *I.M. Drafting*
PROCESSING DEPARTMENT *A. Wafer*   PRODUCT ENGINEERING *H. Yield*
ASSEMBLY DEPARTMENT *W. Bond*   MASK SHOP *J.O. Small*

ENGINEERING CHANGE ORDER COMPLETED _____ DOCUMENT CONTROL DATE _____

FIGURE 1.14 Engineering change orders are a way of life.

Different design groups have different conventions. These are internal preferences and have little effect on the design of the circuit. The convention outlined above is used, as it illustrates the power bus connections to the logic elements.

**1.17
ENGINEERING
CHANGE ORDERS**

One of the biggest problems in all organizations is communication. In small groups, one-to-one verbal conversation may be sufficient. As groups become larger and departmentalized, formal controls need to be instituted to ensure that communication has indeed been effected.

When documentation exists within several independent groups within an engineering organization, an *engineering change order* (ECO) is used to record changes and control the updating of official documents (Figure 1.14). A group designated as *document control* maintains a log of all documents and their revisions. Document control will then monitor and control all changes to the official document.

EXERCISES

1. Name the elements of a MOSFET.

2. What types of MOSFET have a V_P?

3. What types of MOSFET have a V_T?

4. What is a depletion transistor?

5. What is an enhanced transistor?

6. Sketch the side view of a MOSFET.

2

COMMON
LOGIC SYMBOLS

This topic will cover some of the common logic symbols that appear on MOS logic diagrams that you will encounter in industry. The preferred symbols for this course are reviewed in Figures 2.1 to 2.4.

Note that different logic symbols may be needed for grounds. If there are different symbols for ground used on a logic diagram for ground, ask why. In this book, ground is V_{SS} unless otherwise stated, but the three symbols in Figure 2.4 can be used interchangeably for ground. The substitute will be connected to the V_{SS} pin. (Note that although V_{SS} is connected to the V_{SS} ground bus, there will be voltage drops along the bus.)

2.1
INVERTER

The logic symbol for the *inverter* is shown in Figure 2.5a. The schematic for an inverter (Figure 2.5b) may appear in schematic form on the logic diagram. This implementation of an inverter is in ratio logic for the N-channel silicon-gate process.

Sometimes in PLAs (programmable logic arrays) the inverter may appear drawn as shown in Figure 2.5c. Many design engineers will draw the inverter with the inverter bubble on the front of the amplifier symbol (Figure 2.5d).

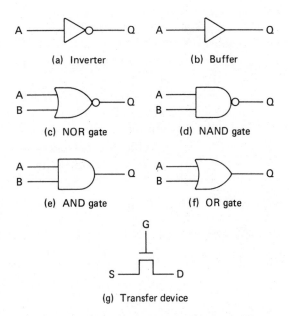

(a) Inverter

(b) Buffer

(c) NOR gate

(d) NAND gate

(e) AND gate

(f) OR gate

(g) Transfer device

FIGURE 2.1 Common logic symbols used for MOS logic.

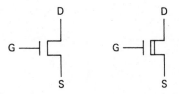

(a) Enhanced transistor   (b) Depletion transistor

FIGURE 2.2 Schematic difference of enhanced transistors and depletion transistors.

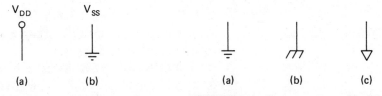

(a)   (b)   (a)   (b)   (c)

FIGURE 2.3 V_{DD} and V_{SS} symbols.   FIGURE 2.4 Common ground symbols.

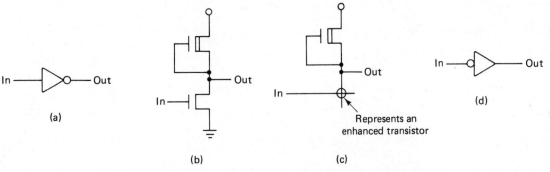

(a)

(b)

(c)

Represents an enhanced transistor

(d)

FIGURE 2.5 The inverter.

2.2
BUFFER

The logic symbol for a *buffer* is shown in Figure 2.6a. The preferred implementation is to use two inverters in series (Figure 2.6b). In this case, the output high level approaches V_{DD} and the output low level approaches V_{SS}.

Another implementation is the *source follower* (Figure 2.6c). This buffer has reduced output high levels. The output high maximum can only be equal to V_{in}-V_T or V_{DD}, whichever is lower. Placing several of these gates in series will cause the output of the last stage to be so poor that the configuration will not function.

In ——▷—— Out

(a)

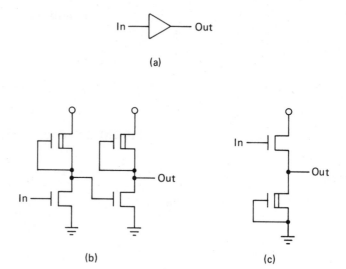

(b)   (c)

FIGURE 2.6   The non-inverting buffer.

2.3
NOR GATE

The *NOR gate* symbol is usually represented as shown in Figure 2.7a or b. However, especially in read-only memories (ROMs), decoders, and PLAs, it may appear as shown in Figure 2.8. Logically, these appear incorrect; however, they are used as a shorthand convenience to represent a ROM array.

Multi-input NOR gates may be represented as in Figure 2.9a or b. The schematic that we will use for a two-input NOR gate will be as shown in Figure 2.10. [Note that part (b) has a NAND shape

In 1 ——|
In 2 ——|⊃o—— Out

(a)

In 1 —o|
In 2 —o|⊃—— Out

(b)   **FIGURE 2.7**   The two-input NOR gate.

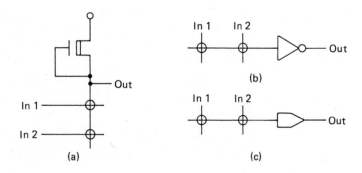

FIGURE 2.8 The two-input NAND gate.

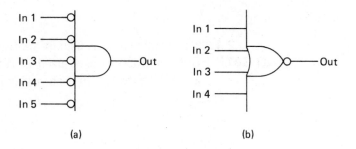

FIGURE 2.9 Multi-input NOR gates.

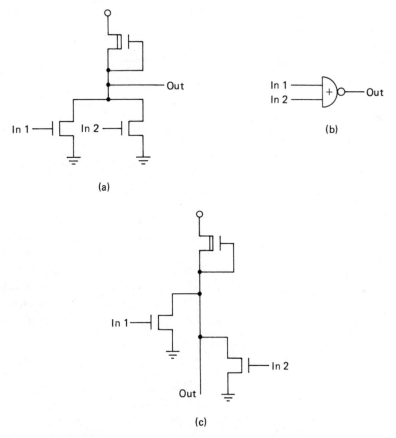

FIGURE 2.10 Common variations of NOR gate used in MOS logic.

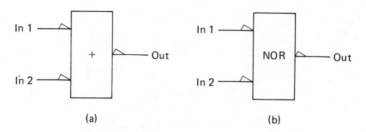

FIGURE 2.11 Other representations of NOR gates.

but is actually a NOR gate.] Sometimes a NOR gate is drawn as shown in Figure 2.11.

2.4
NAND GATE

The *NAND gate* is usually represented as shown in Figure 2.12a or b. Both symbols will be used in this book. The preferred schematic for a NAND gate herein is shown in Figure 2.12c.

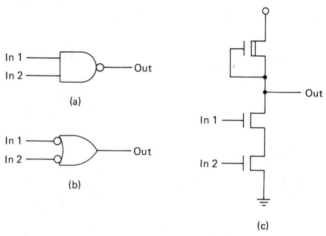

FIGURE 2.12 Common variations of NAND gates used in MOS logic.

2.5
AND GATE

In this book the *AND gate* symbol will be drawn in one of the two ways shown in Figure 2.13. One implementation of an AND gate is a variation of the source follower, using two gates in series (Figure 2.14). This circuit suffers from weak output-voltage high levels.

Figure 2.15 shows the preferred implementation for the AND gate. This circuit has excellent direct-current (dc) output-voltage high and low levels.

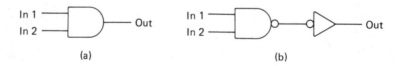

FIGURE 2.13 AND gates.

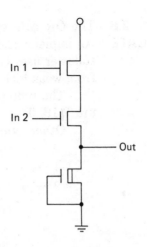

FIGURE 2.14 Single stage AND gate.

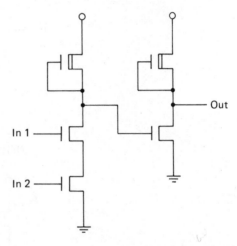

FIGURE 2.15 AND gate formed using NAND and an inverter.

EXERCISE Other symbols you may encounter are shown in Figure 2.16. As an exercise, draw a schematic for each of the four symbols shown.

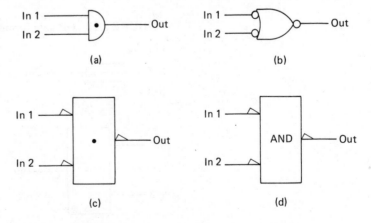

FIGURE 2.16 Other symbols for NAND gate.

2.6   The *OR gate* symbol will be drawn as shown in Figure 2.17a or b.
OR GATE An implementation of the OR gate which is a variation of the source
follower uses two gates in parallel (Figure 2.18). This circuit suffers
from weak high output voltage (V_{OH}).

The preferred implementation for the OR gate is shown in Figure 2.19. This configuration has excellent dc output characteristics.

Other symbols you may encounter are shown in Figure 2.20.

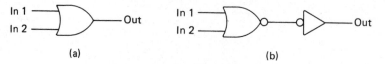

FIGURE 2.17   The OR gate.

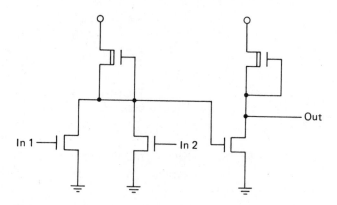

FIGURE 2.18   Single stage OR gate.

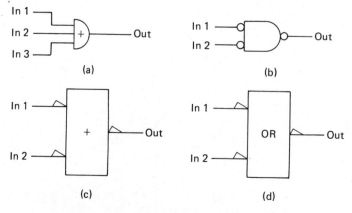

FIGURE 2.19   OR gate formed using NOR and an inverter.

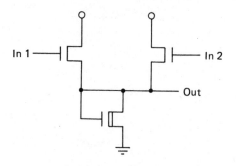

FIGURE 2.20   Other symbols used for OR gate.

Use the design rules and perform the following exercises:

1. Sketch an inverter schematic.

2. Sketch a NOR gate schematic.

3. Sketch a NAND gate schematic.

**2.7
TRANSFER DEVICE
OR COUPLER**
The transfer gate or coupler consists of an enhanced transistor in series with an input or output lead (Figure 2.21a). The device shown in Figure 2.21b is unique to MOS and if properly implemented allows for gating (logically ANDing) and dynamic storage.

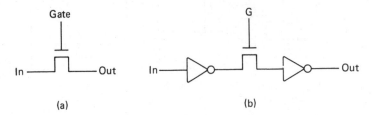

FIGURE 2.21   The coupler.

**2.8
TRANSFER GATE**
The *transfer gate* is more complex and is a CMOS structure (Figure 2.22a). Schematically, it will appear as shown in Figure 2.22b. Sometimes in NMOS the transfer device symbols are used interchangeably on a logic diagram (Figure 2.23).

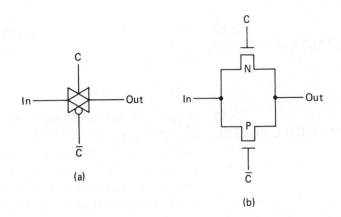

FIGURE 2.22   The transfer gate.

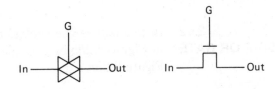

FIGURE 2.23   Symbols used for a coupler.

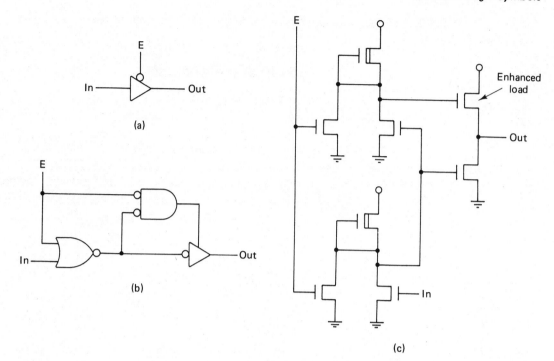

FIGURE 2.24 Three-state buffer.

**2.9
THREE-STATE
OUTPUT BUFFER**

The *three-state* output buffer* is used in NMOS logic parts on multiplexed input/output ports feeding bonding pads. These appear as special symbols and sometimes as schematics. Common representations appear in Figure 2.24.

**2.10
BOOTSTRAP BUFFER**

Another common symbol is the *bootstrap buffer* (Figure 2.25a). The symbol for this may vary. Common schematics are shown in Figure 2.25b–d.

**2.11
PUSH-PULL BUFFER**

A very common logic implementation which requires a special symbol that appears in NMOS is the *push-pull* or *super buffer*. The common logic symbols used are shown in Figure 2.26. Sometimes the symbol in part (a) may have an arrow feeding the output inverter, as shown in Figure 2.26(c). Note that the single amplifier symbol in part (b) is larger than the symbols used in parts (a) and (c).

Schematically, the load for the output is shown sometimes as an enhanced transistor (Figure 2.27a) or sometimes as a combination of enhanced and depletion transistors (Figure 2.27b).

**2.12
EXCLUSIVE-OR GATE**

In this course the *exclusive-OR gate* symbol will be drawn as shown in Figure 2.28a or b. Logically, this may be implemented as shown in Figure 2.28c.

*"Tri-state" is a registered trademark of National Semiconductor Corp. and should not be used in place of "three-state."

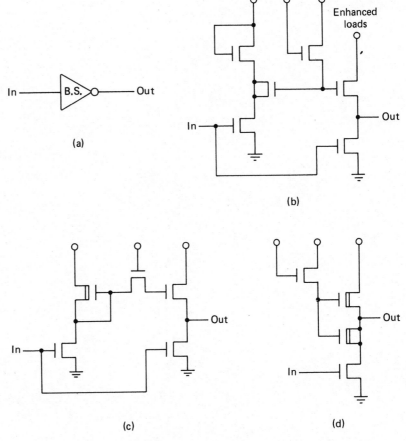

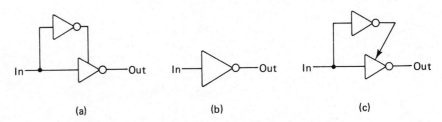

FIGURE 2.25 Bootstrap buffer.

FIGURE 2.26 Push-pull buffer.

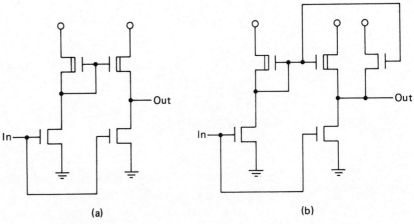

FIGURE 2.27 Schematic drawings for the push-pull buffer.

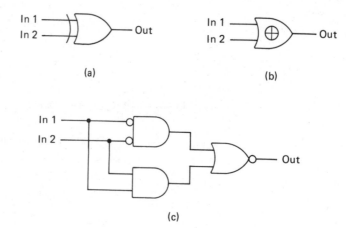

(a)   (b)

(c)

FIGURE 2.28 Exclusive-OR gate.

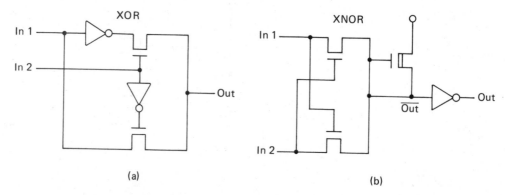

(a)   (b)

FIGURE 2.29 Implementation of XOR gate and XNOR gate.

There are some exotic implementations for *exclusive-NOR* gates. Two examples are depicted in Figure 2.29. The truth table for an exclusive-OR gate is as follows:

In 1	In 2	Out
0	0	0
0	1	1
1	0	1
1	1	0

**2.13
INCLUSIVE-OR GATE**

The *inclusive-OR* gate is nothing more than a standard OR function (Figure 2.30).

**2.14
AND-NOR STRUCTURES**

There are many AND–NOR and OR–NAND functions. An example of an *AND–NOR* appears in Figure 2.31. The thing to remember in creating this type of function is that in positive logic and the N-channel process, *two switches in parallel form an OR gate. Two switches in series form an AND gate.* The load represents the inverter.

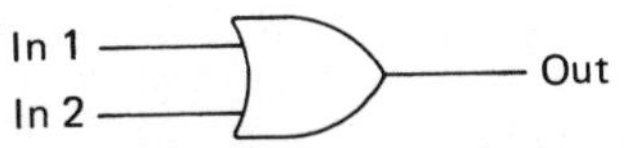

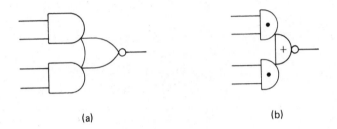

FIGURE 2.30 Two-input OR gate. 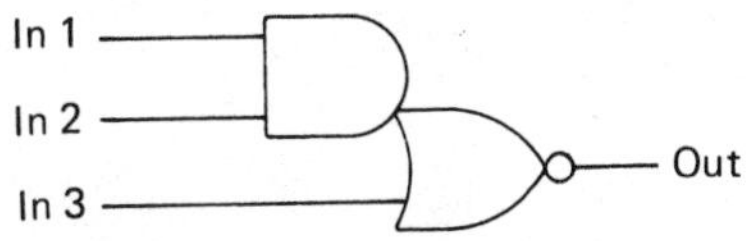   FIGURE 2.31 AND-NOR gate.

**2.15
BUG-EYE**

Occasionally, you may encounter the *"bug-eye"* circuit (Figure 2.32a). It is nothing more than two AND–NOR gates. On a logic diagram it looks something like a bug with two eyes (Figure 2.32b). See Section 3.8.

(a)   (b)

FIGURE 2.32 The "bug-eye"

**2.16
CLOCKING**

To reduce power signal nodes, a dynamic precharge is used rather than a static precharge. This technique is called *clocking*. This is another technique that is unique to MOS. The precharged voltage level is retained by virtue of the capacitance and high impedance of the logic nodes.

Logically clocked inverters are usually drawn as shown in Figure 2.33a and b. Schematics for these appear in Figure 2.33c and d. NOR and NAND gates may similarly be clocked.

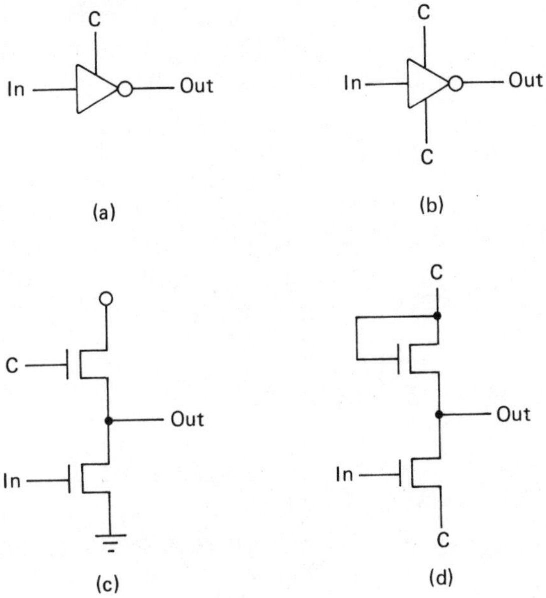

(a)   (b)

(c)   (d)

FIGURE 2.33 Clocked inverter.

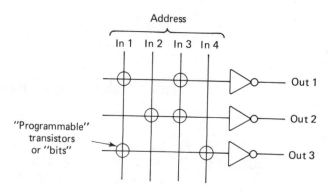

FIGURE 2.34 Programmable logic array.

2.17
ROM
A *ROM* (Figure 2.34) is a read-only memory. ROMs are generally clusters of multi-input NOR gates built in arrays. ROMs are not exclusively NOR gates. NAND gates are sometimes used. Programming ROMs usually means including or removing transistors from the array. This may be done by ion implementation, 06 contact, 01 plugs, 02 mask, 08 mask options, or a combination.

Inputs to a ROM array used as a read-only memory will be an address. ROM arrays may also be used in other ways. Decoders and PLAs are special-purpose ROMs. A decoder is usually a single-level ROM which allows a logic translation of some input. An example is a binary-coded decimal (BCD)-to-Grey code conversion. PLAs are a two-level implementation of a ROM. In this form, an equation containing minimum and maximum terms may be implemented in what logically will be an AND–NOR array.

2.18
RAM
A *RAM* (Figure 2.35), a random access memory, may be static or dynamic. The design engineer will provide details for construction of the various elements.

2.19
LATCHES
AND REGISTERS
A variety of special symbols (Figure 2.36), including those for latches and registers, may appear on the logic diagram. Always question the design engineer as to what these symbols mean. Frequently, a description of the symbol will appear in a cell book or on the logic diagram. Some simple symbols may constitute over one-half of the design.

2.20
OTHER SYMBOLS
Other symbols that may be encountered are shown in Figure 2.37. These are nonstandard symbols. Whenever you encounter a new or nonstandard symbol, ask what it means; never assume that you know.

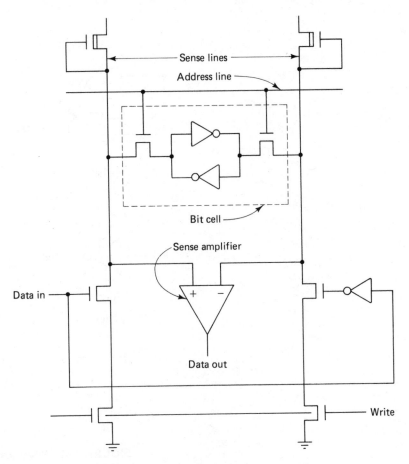

FIGURE 2.35 Static RAM having one bit.

FIGURE 2.36 Common latches.

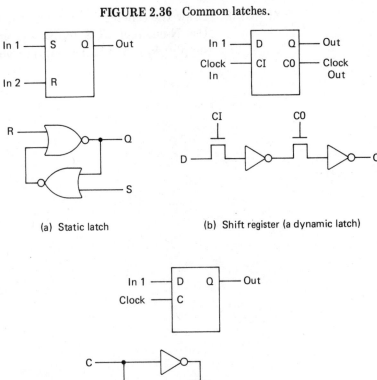

(a) Static latch

(b) Shift register (a dynamic latch)

(c) Static latch

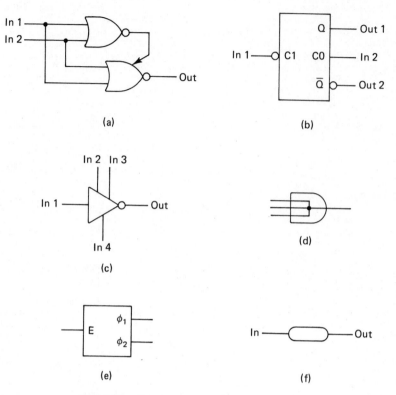

FIGURE 2.37 Not so common symbols.

EXERCISE Draw the schematic diagrams for the symbols shown in Figure 2.37. Use N-channel MOS transistors and positive logic. Assume a single supply: V_{DD} or V_{SS}. Depletion transistors may be used for loads.

3

CONVERTING FROM LOGIC SYMBOLS TO SCHEMATICS

The gates available for implementing digital systems are often NAND gates, NOR gates, or inverters. The choice of using negative or positive logic is normally made by the design engineer. Characteristics of the process, the type of logic, and the power supplies will all be involved in the decision regarding whether to use positive or negative logic. Examples will be given later which should make clearer the problems associated with the choice.

Positive and negative logic can be defined as follows:

Positive logic: Any labeled variable or function is considered to be a logic "1" when the voltage is high. When the variable or function is a logic "0," the voltage is low.

Negative logic: Any labeled function or variable is considered to be a logic "1" when the voltage is low. Conversely, when the function or variable is considered to be a logic "0," the voltage is high.

Thus, in implementing the logical functions with transistors, the choice will be associated with the complexity and difficulty in achieving logic levels. In positive logic using N-channel transistors,

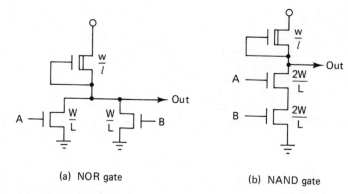

(a) NOR gate   (b) NAND gate

FIGURE 3.1 Positive logic implementation of NOR and NAND gates.

the implementations of NOR and NAND gates in ratio logic are shown in Figure 3.1.*

Following are the truth tables for positive-logic NOR and NAND gates:

NOR Gate Positive Logic				NAND Gate Positive Logic		
A	B	Out		A	B	Out
0	0	+V		0	0	+V
0	+V	0		0	+V	+V
+V	0	0		+V	0	+V
+V	+V	0		+V	+V	0

Note that the two-input NOR gate is easier to implement, as the switches are smaller. Good "1" and "0" levels are easy to achieve for multi-input NOR gates. Additional inputs to the NAND gate require wider and wider switches.

If the logic is to have multi-input NOR gates, the choice for an N-channel process having positive power supplies would be positive logic. Note that if negative logic had been chosen in the parallel structure, a NOR in positive logic becomes a NAND in negative logic (Figure 3.2). The NAND becomes the NOR. Now the logic would

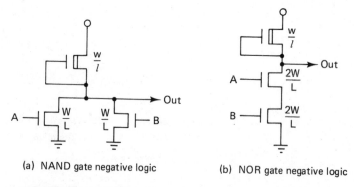

(a) NAND gate negative logic   (b) NOR gate negative logic

FIGURE 3.2 Negative logic implementation of NAND and NOR gates.

*Figure 3.1a is the preferred implementation, as it is approximately one-half the area of the series implementation.

require the use of multi-input NAND gates. The same sort of problem exists in P-channel transistors. Most design engineers like to think that as the power supply goes, so goes the logic levels. V_{SS} equals a logic "0" and V_{DD} is a logic "1."

**3.2
RATIO LOGIC**
Some of the easiest MOS designs to implement use *ratio logic*. Little engineering is required; however, some of the pitfalls need to be understood. Ratio logic relies on the ratio of conductivity of the load to the switch to achieve a low-level output. This ratio and the *aspect ratio* of the transistors are specified by the design engineer.

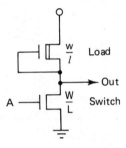

FIGURE 3.3 Elements of an inverter.

To understand ratio and its effects, consider the *ratio inverter* (Figure 3.3). The inverter has two output states: logic "1" and logic "0." When the transistor driven by A is off, the inverter may be considered to be a *resistor* (Figure 3.4a). When the transistor driven by A is on, the inverter may be considered to be a *resistor divider* (Figure 3.4b).

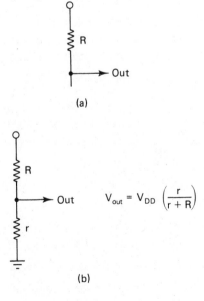

$$V_{out} = V_{DD} \left(\frac{r}{r + R} \right)$$

FIGURE 3.4 A transistor may be represented as a resistor.

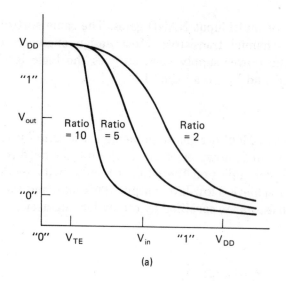
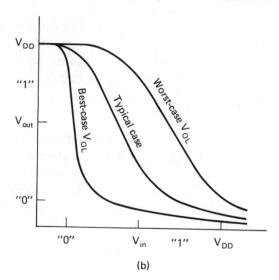

(a)   (b)

FIGURE 3.5   Transfer curves for an inverter.

Determining the ratio requires a definition of logic voltage levels. Normally, the logic low level is chosen to be less than the threshold of the transistor switch. The logic high level is usually picked to be about one-half of the supply level. Design engineers make transfer curves for inverters. This is sometimes done with actual devices from test chips. Often, computer models are used. The input/output voltage transfer curves for three inverters may appear as in Figure 3.5a. Transfer curves are used to determine ratios. Note that the higher the ratio, the greater the gain. The choice of a ratio is complicated because of processing tolerances. This often causes a larger ratio than would be used for typical processing (Figure 3.5b).

For convenience, drawn ratios are specified by the design engineer. To simplify the design, the engineer will specify the switch size to be used with a particular load size. This will be in the form of a device-size table (Table 3.1).

The design engineer is responsible for the correct selection of load sizes for each gate. To determine the load size the engineer will consider such things as propagation time, loading, interconnect, and fan-out. The sourcing or sinking or current must also be considered. Load sizes are computed, simulated, or measured from actual devices on test chips.

TABLE 3.1   Device-Size Table

Size	Load, w/l	NOR, NOT switch, W/L	NAND switch, W/L	Coupler-driven switch, W/L
A	10/25	25/6	50/6	37/6
B	10/15	40/6	80/6	60/6
C	10/10	60/6	120/6	90/6
D	10/6	100/6	200/6	150/6

On test chips, ring relaxation oscillators with buffered test points are used. Figure 3.6 shows an example of a test vehicle for determining propagation delay. Symmetry in layout is used. Several circuits are used for different ratios. A significant value of capacitance is used to simulate loading; 1.0 picofarad (pF) for instance. From this scheme, pair delays may be obtained (Figure 3.7). This information is useful in developing a chart using scaling techniques (Table 3.2). Some design engineers will use the rise and fall time of an inverter driving a capacitor to determine the aspect ratio of a load (Figure 3.8).

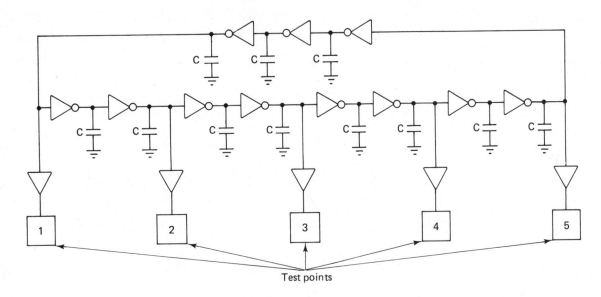

FIGURE 3.6 A ring counter implementation.

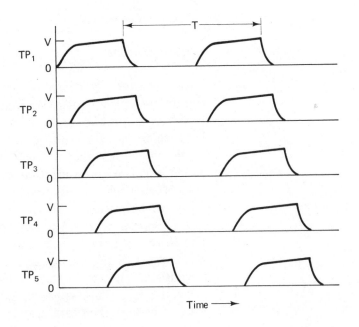

FIGURE 3.7 Wave forms expected from the ring counter.

TABLE 3.2 Aspect Ratio (w/l) for Load Device
for Different Loading and Propagation Times

Loading (pF)	Propagation time (ns)				
	100	200	300	400	500
$\frac{1}{4}$	10/10	10/20	10/30	10/40	10/50
$\frac{1}{2}$	12/6	10/10	10/15	10/20	10/25
$\frac{3}{4}$	18/6	10/7	10/10	10/15	10/16
1	24/6	12/6	10/7	10/10	10/12
$1\frac{1}{4}$	30/6	15/6	10/6	10/8	10/10
$1\frac{1}{2}$	36/6	18/6	12/6	10/6	10/8
$1\frac{3}{4}$	42/6	21/6	14/6	11/6	10/7
2	48/6	24/6	16/6	12/6	10/6

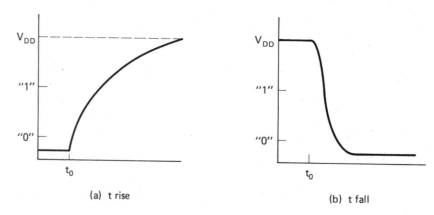

(a) t rise   (b) t fall

FIGURE 3.8 Common rise and fall characteristics of an inverter.

Inverter Current

Ratio inverters, NOR gates, and NAND gates draw current. The amount of current must be considered when figuring power buses. It is very important to keep the IR drops on the main V_{SS} power bus, which supplies voltage to the circuitry, to a minimum. Normally, this is less than 0.2 volt (V). It is the responsibility of the design engineer to ensure that the sizes of the power buses are correct. When back bias is used, the IR drops on the V_{SS} bus are allowed to be greater. The IR drops on V_{DD} can be 0.5 V or more. The IR drops on the V_{SS} bus could bias the source of the switch. The substrate is at 0 V. Body effect will be created on the switch. Body effect is the change in the effective threshold voltage, V_T, of a transistor due to the source being at a different voltage than the substrate, V_{BS}. This will then cause the conductivity, for a certain bias, to be less, therefore destroying the inverter conductivity ratio.

$$V_T = V_{TD} + \tfrac{1}{2}\sqrt{V_{BS}}$$

TABLE 3.3   Inverter Current Requirements[a]

Load size, w/l	Current maximum (mA)	Current minimum (mA)
25/6	1.0	0.2
20/6	0.8	0.16
15/6	0.6	0.12
10/6	0.4	0.08
10/10	0.24	0.048
10/20	0.12	0.024
10/30	0.08	0.012

[a] Ratio = 10, "0" level.

The design engineer will either calculate a model or actually measure a test vehicle to determine currents for different device sizes. The device sizes can be scaled for different currents, generally expressed in milliamperes (mA; see Table 3.3).

To size the power buses properly, an understanding of the different possible logic states is required. For an example, a series of inverters will not all be in the "0" state. Half will be on and half will be off. Most power estimates are based on the following assumptions:

50% of the logic gates are on.

50% of the logic gates are off.

However, knowledge of the circuitry is important in determining actual current or bus lines (Figure 3.9). If required to size buses, consult the design engineer as to the philosophy used or specific cases. Sometimes, large pieces of logic will draw maximum power.

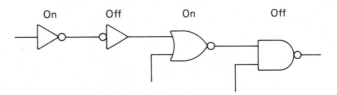

FIGURE 3.9   Effects of logic states.

IR drops. Current through the bus lines will cause *IR* drops (Figure 3.10). *IR* drops on the V_{SS} line will cause gates not to perform as designed. For a switch in the linear region, we find the current to be

$$I_{DS} = \frac{W}{L} K \left[(V_{GS} - V_T) V_{DS} - \frac{1}{2} V_{DS}^2 \right]$$

where I_{DS} = drain-source current
W/L = aspect ratio of the transistor
K = a constant
V_{DS} = drain-source voltage
V_T = effective threshold voltage of the transistor
V_{GS} = gate-to-source bias

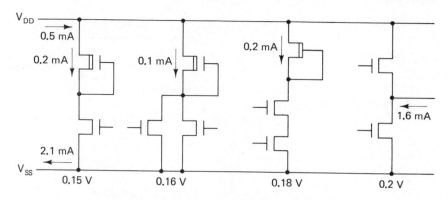

FIGURE 3.10 Voltage drops on a bus line.

In the equation the threshold $V_T = V_{TE} + \frac{1}{2} V_{BS}$, where V_{TE} is the enhancement transistor threshold voltage and V_{BS} is the voltage. Voltage drops on the V_{SS} bus cause V_{BS} to increase. Thus a higher voltage drop on the V_{SS} bus line causes a greater V_{BS}. From the equation above we find that the V_T of the switch transistors will increase. This results in a rise in the low output voltage (V_{OL}) and may cause the gate to fail.

IR drops on buses: limits	*Logic levels internal*
V_{SS}: 0.2 V max.	"0": $V_{SS} - \frac{1}{2} V_{TE}$
	(approx. 0–0.5 V)
V_{DD}: 0.5 V max.	"1": 2 V to V_{DD}

Aspect Ratio

The aspect ratio of a MOSFET is the ratio of its width to its length, and is given by (Figure 3.11)

$$\frac{W}{L}$$

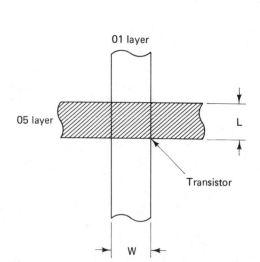

FIGURE 3.11 A transistor as it appears on a drawing.

Only the drawn width and length are given. Design engineers use what are known as *effective widths and lengths*. To determine the effective dimensions, process variations, electrical characteristics, and metallurgical phenomena are used. The effective W/L is necessary to calculate transistor conductivity.

Device Size-to-Device Size Ratio

The device size-to-device size ratio is often referred to as the *beta ratio* or *K ratio*. In this book, the inverter ratio is referred to as the *load-to-switch ratio* or just "ratio." The beta or *K* ratio is a ratio of electrical conductivities. This is useful in determining the output-voltage low levels, but is not necessary for laying out circuits. This ratio is related to the voltage gain of the circuit.

$$\text{aspect ratio} = \frac{W(\text{drawn})}{L(\text{drawn})}$$

W_{eff} and L_{eff} (effective) are used to determine conductivity (Figure 3.12). Drawn device sizes are used. Device size-to-device size ratios are based on drawn sizes and represent the aspect ratios of the switch to the load. An example of ratio is shown in Figure 3.13.

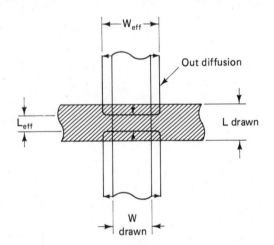

FIGURE 3.12 Effective gates length and width.

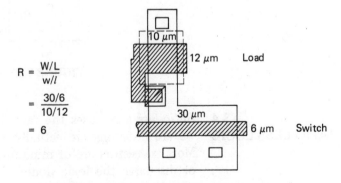

FIGURE 3.13 Inverter ratio is determined by the relationship aspect ratios of the switch and the load.

3.3
CLOCKED AND
PRECHARGED LOGIC

A variation of the ratio logic discussed above is *clocked* or *pre-charged logic*, often using enhanced loads. This technique allows the power dissipation for certain gates to be determined by duty cycle and clock rate. The clock drives the gate of the load, and in some configurations the clock also provides the drain voltage (Figure 3.14). When the load is "on," the output may be precharged to a "1" level, provided that there is no logic "1" level on the switch. This scheme is used with gated clocks. It is also very common in some multiphase clocked logic designs. It is frequently used in conjunction with couplers or transmission devices (Figure 3.15). The capacitors (C_1, C_2) after the transfer devices are necessary to prevent excessive charge removal from the drain switch transistor during the clock transition from "1" to "0." The phenomenon is called *pick off* or *clock decoupling*.

The method of determination of device sizes for clocked logic is not very different from that used to determine device sizes for static ratio logic gates.

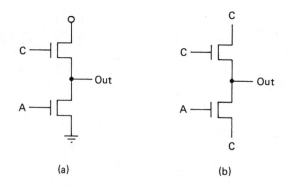

(a)   (b)

FIGURE 3.14   Ratio inverters-clocked.

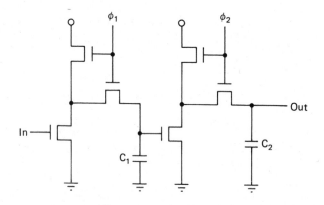

FIGURE 3.15   Clocked shift register.

3.4
RATIO-LESS LOGIC

Ratio-less logic uses the fact that MOS transistors such as the type we are designing are actually capacitors. Each node is a capacitor. Most transistors are of minimum size. Transistors are used to charge or discharge the logic nodes. Just as before, series or parallel transistors can be used to perform logic. Ratio-less logic is used with synchronous systems where there are constant reoccurring clocks

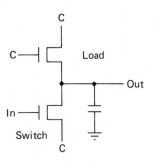

Ratio-less inverter

FIGURE 3.16   Ratio-less inverter.

(Figure 3.16). To work properly, the clocks must have a low output impedance because the capacitance load of the clock lines becomes excessive.

When the clock goes high, the output node is charged high through both the load and the switch. Note in Figure 3.16 that the load and switch are both enhanced. When the clock goes low, the charge remains on the output node unless there is a "1" level on the gate of the switch. A "1" level causes the output node to discharge through the clock line. Other gates may be formed using conventional methods (Figure 3.17).

Ratio-less logic requires only minimum sizes. One of the difficulties with these designs is the need to balance the capacitances of the output node with those of the transistors in the circuit, which requires careful design. One of the advantages is small size, partially because the transistor is of minimum size. Interfacing ratio-less logic with the outside world can be tricky because capacitance balance must be carefully maintained.

Multiphase logic is frequently implemented using ratio-less techniques. A very popular multiphase system uses four phases (Figure 3.18), a *phase* being a clock. One advantage of multiphase logic is that it allows several levels of logic to be accomplished during a clock period.

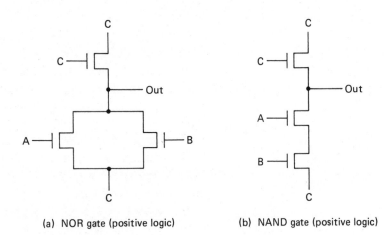

(a) NOR gate (positive logic)   (b) NAND gate (positive logic)

FIGURE 3.17   NOR gate, NAND gate.

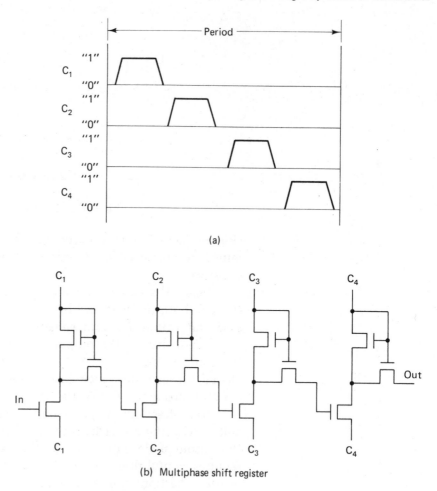

(a)

(b) Multiphase shift register

FIGURE 3.18 Multiphase shift register.

3.5
LOADS: PULLUPS

So far, we have seen that loads may be either static or dynamic. Loads are used to charge nodes to some voltage level. Switches are the transistors that perform the logic on nodes.

Static loads may be either resistors or transistors. The transistors may be of either depletion or enhancement type. (Refer to Section 1.10 for the differences between depletion and enhancement transistors.) The gate of a depletion transistor may be connected by its source or drains to a special supply, or to a logic node, such as another gate. The gate of the enhanced transistor may be connected to either a special supply or to its own drain supply. It also might be connected to another logic element, as in the case of a push-pull buffer.

Dynamic loads are almost always enhanced. These gates are usually connected to some clocking logic. This type of load is used to precharge specific nodes. A load connected to the V_{DD} supply is sometimes referred to as a *pullup*.

Examples of loads are shown in Figure 3.19. There are many variations; therefore, the design engineer must always specify what type of load to use. Frequently, the type of load specified appears

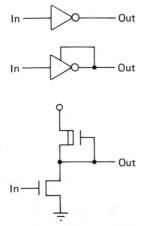

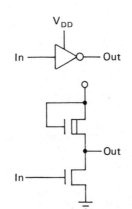

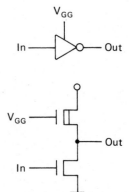

(a) Static depletion load gate connected to source

(b) Static depletion load gate connected to V_{DD}

(c) Static depletion load gate connected to V_{GG}

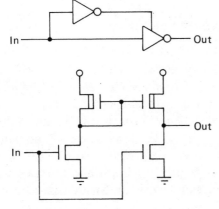

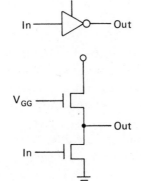

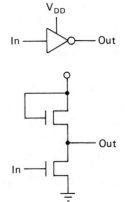

(d) Static depletion load gate connected to a logic element

(e) Static enhanced load gate connected to V_{GG}

(f) Static enhanced load gate connected to V_{DD}

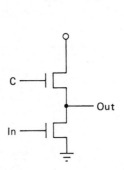

(g) Dynamic enhanced load gate and switch drain connected to C

(h) Dynamic enhanced load gate connected to C

FIGURE 3.19 Inverter implementation.

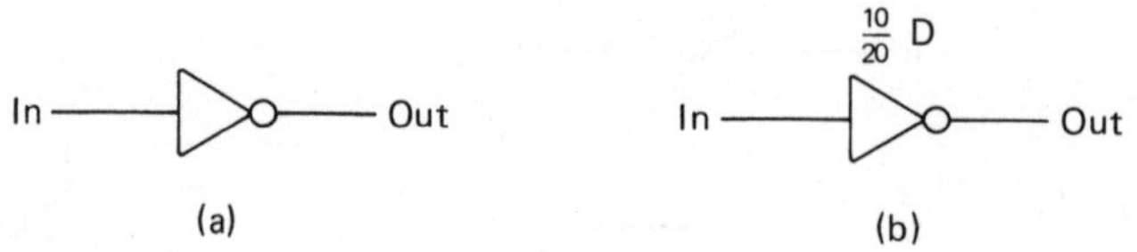

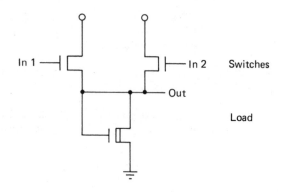

FIGURE 3.20 Method of specifying device sizes and type.

FIGURE 3.21 OR gate.

as a general note on the logic diagram. Load sizes are specified by the design engineer. Sometimes the load sizes are specified by an encompassing note. Frequently, every gate is specified. The load size appears at the output of the gate as illustrated in Figure 3.20b. This specifies the aspect ratio (W/L) of the load to be of the dimensions and type called for in the device-size chart. The 10/20 D specifies that the load have a drawn width of 10 μm and a drawn length of 20 μm, and that it be a depletion transistor. Switch sizes are specified in the device-size charts.

Rarely, logic is performed with *load devices*. When this is done, the load takes on the position of a source follower (Figure 3.21). These loads are enhanced.

3.6 SWITCHES: PULLDOWNS

The switches or *pulldowns* or, in special applications, killers and prechargers, are used to perform logic on nodes. A *killer* is a switch in which the drain is connected to a logic node and the source is tied to V_{SS}. This will ground the node to V_{SS} when the gate voltage goes high. A *precharge* is a switch in which the drain is connected to V_{DD} and the source is connected to a logic node. When the gate on the precharge goes high, the logic node is clamped to V_{DD}. Switches are almost always enhanced. The gates of switches are normally connected to a logic element. Occasionally, a switch is tied to a supply. Switch sizes are specified either in the device-size chart on the logic diagram as a special note or at the input of the gate. They are specified by the design engineer.

Switches usually have shorter channel lengths than depletion load devices. There are many reasons for this, some design and some electrical. Switches may be connected in series or in parallel.

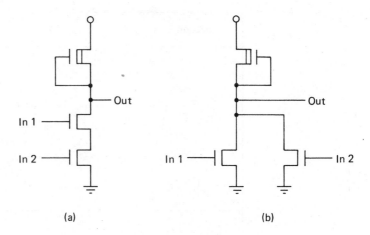

FIGURE 3.22   NOR and NAND gates.

**3.7
PARALLEL AND
SERIES TRANSISTORS**

Switches are connected in parallel and series to perform logic on nodes (Figure 3.22). We have seen how parallel transistors and series transistors can form NOR and NAND functions.

To obtain the same current capability on a node, two switches in series have their width doubled. Switches driven by couplers, as a general rule, are made 50% larger or sometimes doubled to compensate for the loss of drive after a coupler due to pickoff. A coupler is a transistor that is used in series between two nodes to perform logic.

**3.8
DeMORGAN'S THEOREM
AND BOOLEAN
ALGEBRA LAWS**

A technique that is frequently used by design engineers for converting logic is the use of *DeMorgan's theorem*. It allows conversion from AND to OR functions. DeMorgan's theorem is also known as the *law of dualization*.

DeMorgan's Theorem

To perform DeMorgan transformation:

1. Complement the entire expression.
2. Change "+" to "·" and "·" to "+."
3. Complement each variable.

That is,

$\overline{0} = 1$

$\overline{1} = 0$

"·" means "AND"

"+" means "OR"

Figure 3.23 shows the logic symbols and circuit diagrams for the following examples.

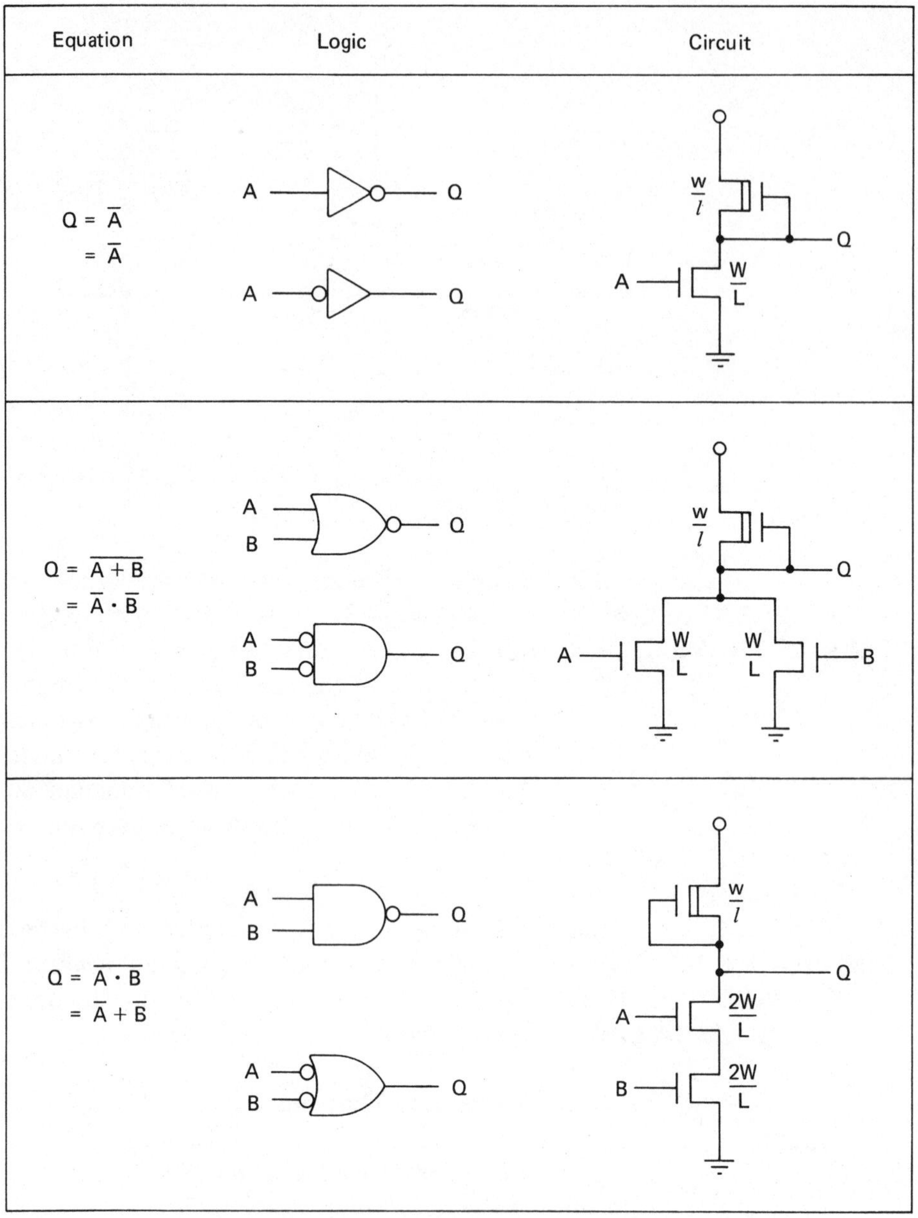

FIGURE 3.23 Common MOS logic symbols and circuit diagrams illustrating DeMorgan's theorem.

$$A \cdot B = \overline{\overline{A} + \overline{B}}$$
$$\overline{A \cdot B} = \overline{A} + \overline{B}$$
$$\overline{A + B} = \overline{A} \cdot \overline{B}$$

Boolean Algebra Laws

Following are some useful Boolean algebra laws.

Absorption law
$$A + A \cdot B = A$$
$$A(A + B) = A$$

Quantities that are parts of whole quantities may be combined with the whole quantity.

Associative law

$$(A + B) + C = A + (B + C)$$
$$(A \cdot B) \cdot C = A \cdot (B \cdot C)$$

Quantities in groups may be reordered without affecting the result.

Law of double negation

$$\overline{\overline{A}} = A$$

The complement of the complement of a quantity is the equivalent of the quantity.

Commutative law

$$A + B = B + A$$
$$A \cdot B = B \cdot A$$

Changing the order of the terms in an equation will not affect the value of the equation.

Complement law

$$\overline{A} \cdot A = 0$$
$$\overline{A} + A = 1$$

The addition of a quantity and its complement will result in a sum of 1 and the multiplication of a quantity and its complement will result in 0.

Distributive law

$$A \cdot (B + C) = A \cdot B + A \cdot C$$
$$A + (B \cdot C) = (A + B) \cdot (A + C)$$

Quantities in groups operated on by operators may be separated from the group with the appropriate operation retained.

Idempotent law

$$A + A = A$$
$$A \cdot A = A$$
$$0 + 0 = 0$$
$$0 \cdot 0 = 0$$
$$1 + 1 = 1$$
$$1 \cdot 1 = 1$$

Combining a quantity with itself will result in a quantity equivalent to the original quantity.

Identity law

$$A = A$$

A quantity is equivalent to itself.

Axioms
1. $0 + A = A$
2. $1 + A = 1$
3. $0 \cdot A = 0$
4. $1 \cdot A = A$
   $0 \cdot 1 = 0$
   $0 + 1 = 1$

3.9
COMPLEX GATES

Complex gates are frequently represented by a box. The logic that appears in the box is defined on cell sheets or on the logic diagram by the design engineer. Logic represented by a box may be nothing more than a single transistor or it may be a very complex set of logic. Always question the contents of undefined, unfamiliar logic symbols on a logic diagram. Figure 3.24 illustrates more complex MOS logic symbols and circuit diagrams.

Bubble Game

Logic can be reduced in a variety of ways using various techniques. This book was not set up to teach logic design. However, there are some techniques that may be used to assist in a practical way in reducing logic. Before any changes are made to the logic, consult with the design engineer.

The bubble game is one useful means of logic conversion. A bubble in NMOS ratio logic implies a load. It is the symbol for logical inversion. Two bubbles in series mean no inversion. A bubble may be placed at the front or rear of an inverter. A NOR gate may be drawn as a negative-input AND gate. A NAND gate may be drawn as a negative-input OR gate. Figure 3.25 shows examples of these procedures. If we then move the bubble around and combine bubbles, we can reduce logic (Figure 3.26).

3.10
CMOS

CMOS (complementary metal-oxide semiconductor) gates are introduced in this text for information only. The CMOS processes have the advantages of having both P- and N-channel transistors. Enhanced and depletion transistors are frequently available. Older processes had only enhanced transistors.

The CMOS inverter consists of an N-channel and a P-channel transistor tied in series with their gates tied to a common input (Figure 3.27). The common node between the N- and P-channel transistor is the output. Logic frequently appears in both N-channel and P-channel transistors. The biggest advantage with CMOS logic designed as shown in Figure 3.27 is that it is static and does not draw power in its stable states.

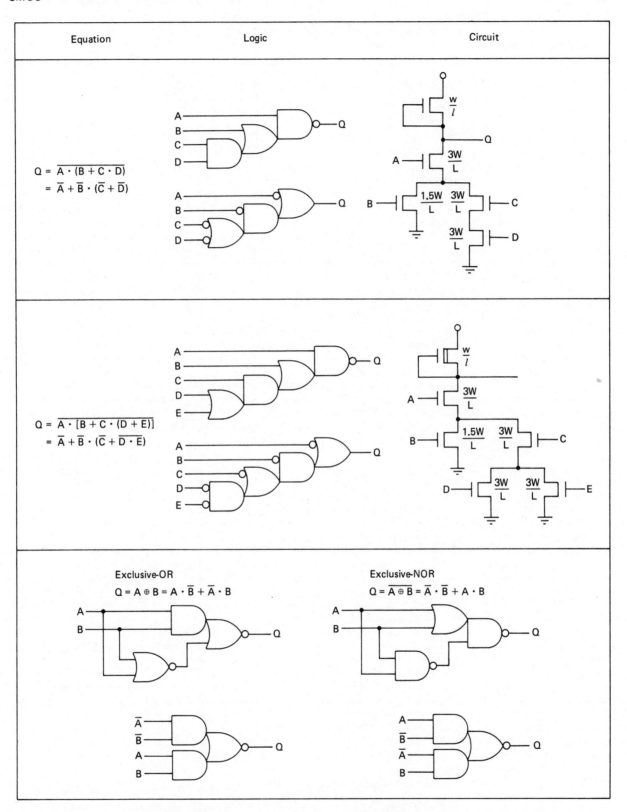

FIGURE 3.24 More complex MOS logic symbols and circuit diagrams.

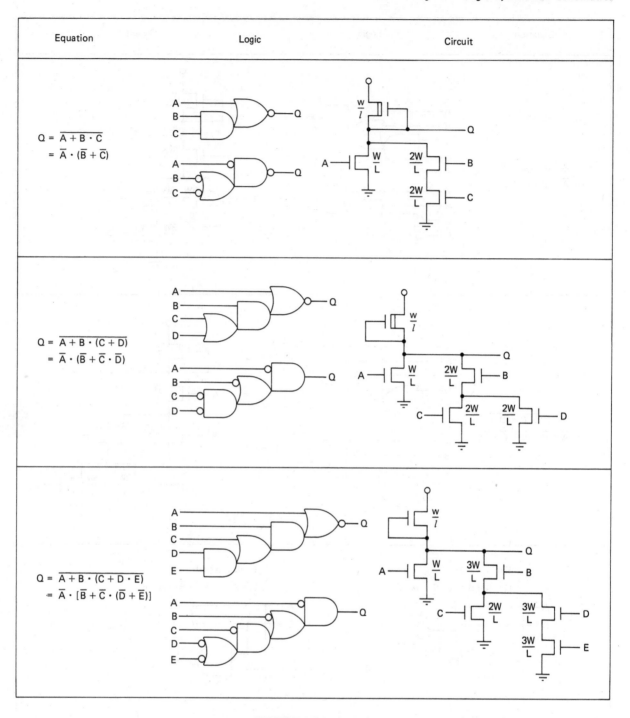

FIGURE 3.24   (cont.)

Current trends in MOS today are toward a high-density CMOS process. CMOS logic takes more area to implement than does a single-channel process. As systems become more complex, this difference becomes less and less. This is because interconnections take up more and more of the chip area. As the parts become more com-

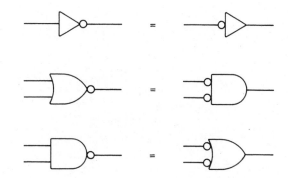

FIGURE 3.25 Equivalent symbols.

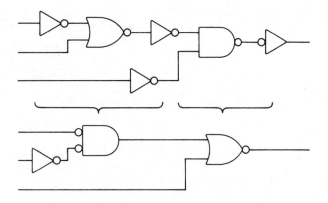

FIGURE 3.26 Logic simplified by using equivalent symbols.

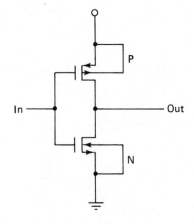

FIGURE 3.27 Schematic for CMOS inverter.

plex, the power becomes greater. Too much power may exceed package capabilities. The CMOS logic offers a solution to these problems. The high-density processes allow more gates in the same area (Figure 3.28).

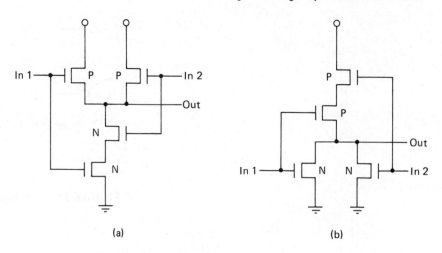

(a)   (b)

FIGURE 3.28 Schematic drawing for CMOS (a)AND gate and (b)CMOS NOR gate.

EXERCISES   Draw a schematic for each of the following circuits.

1. $Q = \overline{A}$

2. $Q = \overline{A} \cdot \overline{B}$

4

DIODES, INTERCONNECTIONS, AND CONTACTS

The nature of semiconductors is such that whenever N-doped material comes in contact with P-doped material, a diode is formed. If the materials are both heavily doped, little diode action may be noted. The material will conduct easily in both directions. Good diode action is achieved if one material is lightly doped and the other is heavily doped (Figure 4.1). The large concentration of either holes or electrons in the heavily doped material will cause a large depletion region to exist in the lightly doped area. This creates good diode characteristics. Thus the presence of heavily doped N^+ regions in lightly doped P^- substrate creates excellent diodes. These diode regions are used for sources and drains for transistors and for interconnections. General diode characteristics are shown in Fig. 4.2.

The N^+ region has resistance. Because the depth is controlled within certain tolerances by processing, the resistance of the N^+ regions may be expressed in terms of *sheet resistance* (sheet rho or ρ). A study of the diode junction reveals a depletion region. In this area the necessary ingredient for capacitance exists: two electrodes and a dielectric. Capacitance is expressed as capacitance per area and side-wall length. The polycrystalline silicon and metal layers are insulated from other layers and the substrate except where contact is made. The resistance in these layers is expressed in terms of sheet resistance. The capacitance is expressed as capacitance per unit area.

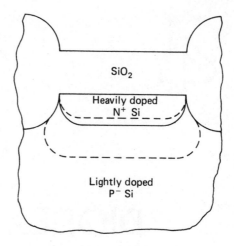

FIGURE 4.1 Cross section of a diffusion line showing the N+ P— diode and the diode depletion region.

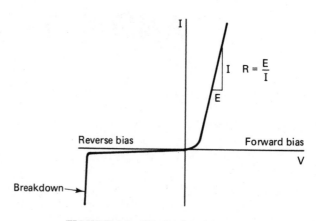

FIGURE 4.2 Diode characteristics.

4.1 RESISTORS Resistors resist the flow of current. The current that can pass through a resistor is given by Ohm's law:

$$I = E/R$$

where I = current
E = potential applied across the resistor
R = resistance

In the N-channel silicon-gate process, the N^+ regions, polycrystalline silicon regions, and metal regions all have resistance. The range of values for the resistance are specified in the process definition and limits. For our process, the following ranges are assumed, in units of ohms per square ($\Omega/\square$):

N^+	10 to 20
Polycrystalline silicon	20 to 60
Metal	0.01 to 0.03

These values allow for process tolerances due to thickness width and length. The element used to monitor line width is some object on a chip common to other chips. Usually, this is a line on a process monitor test site. The tolerances or critical dimensions are referred to as *mask critical dimensions* and *after-etch critical dimensions*.

Note that the resistance is specified in ohms per square, drawn. This allows a designer to determine the resistance of a line simply by breaking the shape into squares (Figure 4.3). Corners are assumed to be one-third of a square.

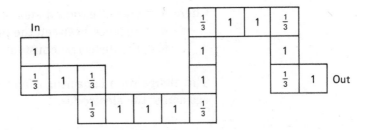

FIGURE 4.3   A total of $13\frac{1}{3}$ squares.

Because all lines contain some resistance, it is necessary to set a routing priority. Generally, V_{SS} has the highest priority. Next come the clock lines and other critical nodes. Next is V_{DD}. These lines are routed on metal wherever possible. Signal lines are frequently routed on polycrystalline silicon and N$^+$ for short distances, and metal for longer distances. Where lines of the same type come into conflict, such as two lines perpendicular to each other, tunnels or crossovers are used. Crossovers and tunnels are similar to "feedthroughs" on printed circuit boards. One of the signal lines in conflict will be routed to a different layer through contacts.

**4.2
CAPACITORS**

To create a capacitor, two electrodes and a dielectric material are required. The electrodes need not be conductive, but must hold a charge. The transistors and the interconnect lines of a MOSFET IC are all capacitors. They all have at least two electrodes and some sort of dielectric. In the case of the transistor gate, the electrodes are the gate, source, drain, and bulk, and the dielectric is the silicon diode. In the case of the interconnections, the electrodes comprise the interconnect line and the substrate or another interconnect line or transistor gate. The dielectric is the silicon diode. In the case of the transistor source and drain and N$^+$ interconnect, one of the electrodes is the N$^+$ region, the other is the substrate. The dielectric is the depletion region that exists in the silicon substrate.

Physics tells us that the capacitance between two flat plates may be described as the ability to hold a charge. Capacitance varies directly as the exposed area of the two plates and inversely proportional to the distance between the plates.

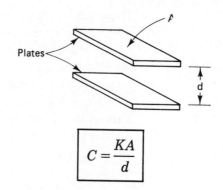

$$C = \frac{KA}{d}$$

where A = area of exposed area of plate
$\quad\quad d$ = distance between the plates
$\quad\quad K$ = the dielectric constant

The charge on a capacitor is equal to the capacitance times the voltage across the electrodes.

$$Q = CV$$

where Q = charge in coulombs
$\quad\quad C$ = capacitance in farads
$\quad\quad V$ = voltage difference between the electrodes in volts

The effective value of capacitance of two or more capacitors in series is given by

$$\frac{1}{C} = \frac{1}{C_1} + \cdots + \frac{1}{C_N}$$

where C is the effective capacitance and C_1 through C_N are the different capacitor values in series.

The effective value of capacitance of two or more capacitors in parallel is given by

$$C = C_1 + \cdots + C_N$$

The properties of capacitors are such that it is the electric field created by the voltage difference between the electrodes that makes the capacitor work (Figure 4.4a). The electric field fringes are located around the edges of the plate (Figure 4.4b). The fringing of the electric field will alter the value of capacitance, especially for small conductors. Since all the minimum-size polycrystalline silicon and metal interconnects manifest fringing, this effect must be taken into account when fringing the capacitance for these lines. Rather than attempting to calculate the capacitance from the area values,

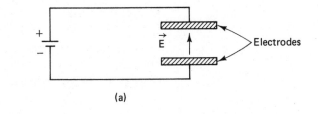

(a)

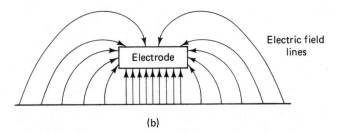

(b)

FIGURE 4.4 (a) Battery connected to two parallel plates acting as electrodes and the resulting electric field; (b) gate capacitance effects electric field lines of a rectangular electrode over a flat surface.

the following values will be used in this book when calculating interconnect capacitance:

Polycrystalline silicon routing at 6 μm (6 μm wide)	0.56 fF/μm
Metal routing at 8 μm	0.28 fF/μm

Note: fF = femtofarad = 10^{-15} farad.

The characteristics of the capacitance with respect to the voltage applied vary with the electrodes used (Figure 4.5). Gate capacitances vary within the operating voltage range from a maximum to

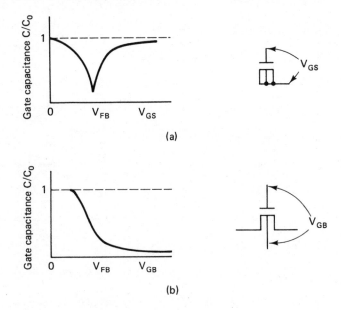

(a)

(b)

FIGURE 4.5 Gate capacitance effects.

a minimum and back to a maximum. Metal and polycrystalline silicon interconnect capacitance values also vary with voltage, but remain practically constant over the normal operating voltage. The capacitance in N^+ regions has the same characteristics as in diodes (Figure 4.6). This value of capacitance drops with the voltage applied.

When calculating N^+-region capacitance, it is necessary to consider three components (Figure 4.7):

1. C_{area}, the area exposed to the bulk
2. C_{SWF}, the side wall exposed to the field
3. C_{SWG}, the side wall exposed to the gate region

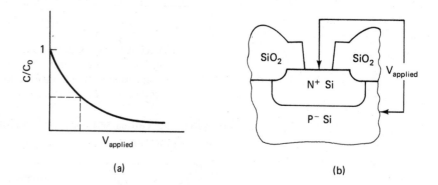

FIGURE 4.6   Junction capacitance characteristics.

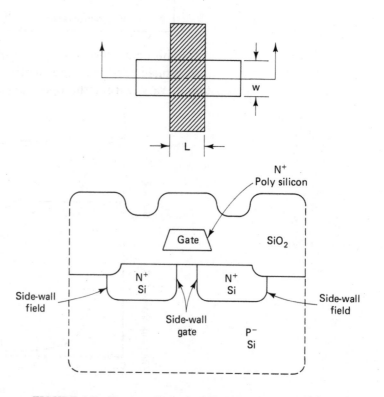

FIGURE 4.7   Cross section of a MOS N-channel transistor.

4.3
SIGNAL BUSES

Signal buses are interconnect lines with a large fan-out. Signal buses usually originate internally from logical gates. Sometimes the bus signals are directly connected. To improve performance, it is necessary to keep loading to a minimum. It is most desirable to route any signal lines via a route with the least amount of capacitance. The best way to do this is to prepare a good plan and keep the signal lines on metal. Partition and organize the logic to reduce the length of interconnects. In microprocessor and microprocessor-type design, groups of signals are routed together. These might be address or data buses. These signals usually have a high routing priority and cross-unders are kept to a minimum.

4.4
INTERCONNECTIONS

Normal interconnections in cells and between cells take the lowest priority in routing. These signals may occur on N$^+$, polycrystalline silicon, or metal (Table 4.1). Certain nodes that are sensitive to excess capacitance should be specified by the design engineer. The first inverter of an input buffer is an example. Most interconnect signals have a small fan-out.

4.5
POWER BUSES

Power buses should be routed on metal because of the current requirements. Some branches going to a small group of gates may be on an N$^+$ or polycrystalline silicon cross-under or tunnel. On almost all ratio logic designs the power buses should be designed for minimum voltage (IR) drops. In this book the maximum voltage drop allowed on any power bus is *0.2 V.* Ion migration is a consideration

TABLE 4.1 Interconnect Capacitance and Resistance

Nodal capacitance	
N$^+$ bottom	0.070–0.150 fF/μm^2
N$^+$ side-wall field	0.500–0.800 fF/μm^2
N$^+$ side-wall gate	0.250–0.400 fF/μm^2
Polycrystalline silicon over gate	0.300–0.350 fF/μm^2
Polycrystalline silicon over field	0.030–0.060 fF/μm^2
Metal over field	0.018–0.036 fF/μm^2
Metal over N$^+$ or polycrystalline silicon	0.033–0.066 fF/μm^2
Routing capacitance	
N$^+$ 8 μm wide	0.400 fF/μm
Polycrystalline silicon 6 μm wide	0.560 fF/μm
Metal 8 μm wide	0.280 fF/μm
Nodal resistance	
N$^+$ sheet	10–20 Ω/$\square$
Polycrystalline silicon sheet	20–60 Ω/$\square$
Metal sheet	0.01–0.03 Ω/$\square$
Routing resistance	
N$^+$ 8 μm wide	1.12–4.16 Ω/μm
Polycrystalline silicon 6 μm wide	3.00–14.8 Ω/μm
Metal 8 μm wide	0

in sizing power buses. Ion migration may occur when the current density exceeds 1 mA/μm at temperatures in excess of 150°C. At these elevated temperatures and currents the metal traces on a circuit will open and break.

4.6 TUNNELS AND CROSSOVERS

Signals and power on metal may be routed under other metal lines by using an N$^+$ or polycrystalline silicon cross-under. Where power is a concern, N$^+$ is preferred. On signal buses, polycrystalline silicon is preferred. Sometimes a combination of N$^+$ and polycrystalline silicon may be used. This might be where space is a concern.

Polycrystalline silicon lines may use metal crossovers. N$^+$ may use metal crossovers. Where resistance and capacitance is critical, the time constants must be calculated.

4.7 TIME CONSTANTS

If the time constant of a line is too large, no matter how the gate W/L is sized, the signal will not propagate in sufficient time. The *time constant* is a relation of the resistance times the capacitance.

In some arrays, certain gating signals are routed in polycrystalline silicon and their time constants may be too large, resulting in logic race conditions. Killers or precharges might be used at the ends of the long polycrystalline runs to correct the race conditions.

Time constants are a major concern of design engineers.

4.8 CURRENT DENSITY

If the junction temperature T_j of a chip reaches 150°C and the current density J of an aluminum line reaches 10^6 A/cm^2, a strange phenomenon occurs. The line will disintegrate in slow motion. At $J = 10^7$ A/cm^2, the line will fuse.

Line disintegration in slow motion is called *ion migration*. To prevent ion migration, current density for metal is set at 25 mA/mil or 1 mA/μm. Ion migration is not a problem with deposited polycrystalline silicon interconnect lines since resistance of the lines limits the current flow; therefore, the current densities required for ion migration do not exist. Fusing does occur due to the heat resulting from current flow and resistance.

4.9 LOADING

Capacitance is a major contribution to the loading of a MOSFET gate. Most of the nodes drive interconnect and transistor gates. Sometimes, for special logic elements, the loading on a gate includes current. The MOS cheater latches are an example. Output buffers are prime examples, as they must "sink" and "source" TTL values of current. It is important that every MOSFET IC designer consider loading when laying out a mask design. Most loading should be kept to a minimum. There are special cases in certain delay circuitry where loading is made a specific value, such as in a "one-shot" circuit.

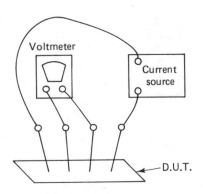

FIGURE 4.8 Method for measuring sheet resistance.

4.10
SHEET RESISTANCE

As already indicated, resistance for N^+, polycrystalline silicon, and metal is specified as sheet resistance in terms of ohms per square. Sheet resistance is normally measured with a four-point probe (Figure 4.8). To determine the resistance of a line, the effective number of squares that exist in the line are calculated.

Example. A polycrystalline silicon line of 120 μm and 6 μm wide will have 20 squares and in the worst case will have 1200Ω. The values for sheet resistance for N^+, polycrystalline silicon, and metal are given in Section 4.1.

4.11
CONTACTS

In this process there are two types of contacts available:

1. The buried or 03 contact
2. The preohmic metal or 06 contact

These contacts act similar to the way in which feedthroughs act on a printed-circuit board, and permit contact between layers. A contact allows for connection between conducting layers separated by the oxide insulator.

The 03 contact or buried contact removes the gate oxide from beneath the polycrystalline silicon lines. If there is an N^+ region, contact between the N^+ region and polycrystalline silicon is made. The preohmic or metal contacts define where the field oxide will be cut to allow connection between the metal layer with the N^+ and 05 polycrystalline silicon regions.

5

MOS TRANSISTORS

The elements that make up the electrodes of the MOS transistor are the source, drain, gate, and substrate bulk. The insulated-gate field-effect transistor is a capacitor. One electrode of the capacitance is the gate. The other electrode consists of the source, drain, and bulk. There exists a thin film of silicon dioxide, SiO_2, 1000 Å thick between the gate and substrate.

5.1 ENHANCED TRANSISTOR

The enhanced transistor is built on the surface of a lightly doped P-type silicon substrate. Two N-doped regions 12,000 Å deep form the source and drain. A thin film (1000 Å thick) of SiO_2 lies on the surface of the silicon between the source and drain. The gate consists of N-doped deposited polycrystalline silicon 4000 Å thick. The silicon substrate has been lightly doped by ion implantation to change its work function.

The N^+ regions in the P^- region form a diode. Around the metallurgical junctions a depletion region is formed. A depletion region is devoid of free holes or electrons. The work-function difference between the regions establishes an electric field that drives the free holes or electrons from the region. If the reverse bias of the N^+ region increases, the depletion region becomes larger. If the N^+ region is forwarded-biased, the field is overcome and there is conduction between the N^+/P^- junction.

Because the N⁺ regions in the P⁻ substrate make good diodes, N⁺ regions are used as electrodes for sources, drains, and interconnects. To provide conduction between the source and drain, the gate electrode is biased with a positive voltage. The electric field set up between the gate and the substrate draws electrons to the surface. The excess of electrons changes the P material to an electrically neutral state (referred to as *intrinsic*). Additional gate voltage creates a stronger field, drawing more electrons to the surface and causing the material to appear as N-type material. If voltage is applied between the source and drain at this time, conduction will occur.

On the composite drawing, a transistor will appear as the intersection of the 01 region and the 05 region (Figure 5.1a). The gate bias at which the energy bands at the substrate silicon dioxide interface just balance is called the *flat-band voltage*. The gate bias at which conduction takes place between the N⁺ regions is known as the *threshold voltage*.

Applying a voltage between the source and drain will disturb the depletion regions and alter the threshold voltage. The depletion region around the source sets up a field to counter the flow of current from the drain. If the source is biased positive with the substrate, the source depletion region becomes larger, further impeding the flow of current. To offset this field, more gate voltage must be applied. The effect on the threshold voltage is given by the relationship

$$\Delta V_T = \tfrac{1}{2}\sqrt{V_{BS}}$$

where ΔV_T is the change in threshold voltage and V_{BS} is the bias on the source N⁺ region with respect to the substrate (or bulk).

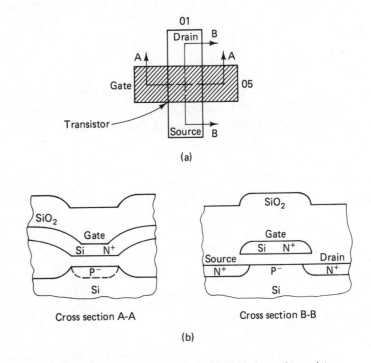

(a)

(b)

FIGURE 5.1 Cross sections of a MOS N-channel transistor.

The depletion region around the drain N⁺ region increases with an increase in bias. Electrons that enter this region are rapidly accelerated across the region. The bias on the source affects the threshold voltage. For long channel transistors the drain depletion region has little effect on the current. As the channels become shorter, the effects of the depletion regions become stronger.

The effective channel length of the transistor is that region of the substrate surface between the source and drain depletion regions. This can be determined if the size of the regions can be predicted.

5.2
NARROW CHANNEL
TRANSISTORS

The fringes of the transistors near the field are affected by the field implant (Figure 5.2). The field implant inhibits the transistor inversion. As a result, the threshold voltage for narrow transistors is higher than for wide ones.

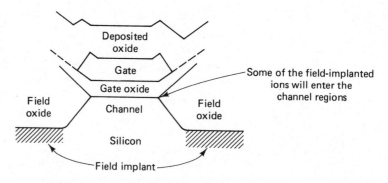

FIGURE 5.2 Field implant encroachment into the gate region of a MOS transistor.

5.3
SHORT CHANNEL
TRANSISTORS

Short channel transistors tend to act more like gated depletion regions (Figure 5.3). Weak inversion currents exist. Even when the gate bias is less than the threshold voltage, current still flows.

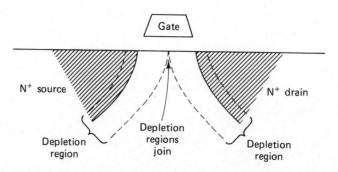

FIGURE 5.3 Short channel effects of MOS transistor are caused by depletion regions.

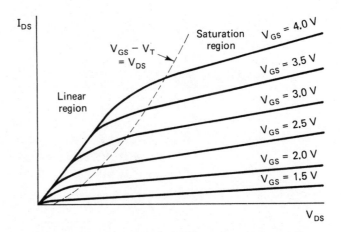

FIGURE 5.4 I_{DS} vs V_{DS} characteristics of an enhancement NMOS transistor.

5.4
MOS TRANSISTOR MATHEMATICAL MODEL

In 1964, C. T. Sah* published two equations to predict the drain-source current through large circular MOS transistors. One equation is for the source drain current nonsaturation region, and the other is for the source drain current saturation region. From his work most of the mathematical models for MOS transistors have been derived.

The characteristic I_{DS} versus V_{DD} curves for a typical MOS transistor are illustrated in Figure 5.4. For the purposes of this text the characteristics for the MOS transistor may be described with the following equations:

1. Nonsaturation-region equation:

$$I_{DS} = K\frac{w}{l}\left[V_{DS}(V_{GS} - V_T) - \frac{V_{DS}^2}{2}\right]$$

2. Saturation-region equation:

$$I_{DS} = K\frac{w}{l}(V_{DS} - V_T)^2$$

Here I_{DS} = drain–source current
w/l = aspect ratio of the transistor
K = a constant
V_{DS} = drain–source voltage
V_T = effective threshold voltage of the transistor
V_{GS} = gate-to-source bias

Note: When $V_{GS} - V_T = V_{DS}$, the two equations have equal values for I_{DS}. Figure 5.5 is a schematic depicting the voltages used in the equations.

*The equations appeared in an article by C. T. Sah, "Characteristics of the Metal-Oxide-Semiconductor Transistor," *IEEE Trans. Electron Devices*, pp. 324–345, July 1964.

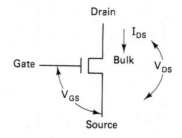

FIGURE 5.5 Schematic of MOS transistor showing V_{GS} and V_{DS}.

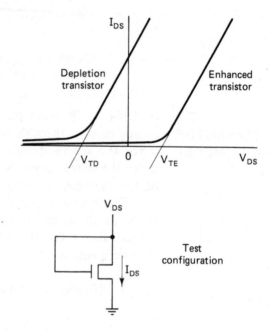

FIGURE 5.6 Method and difference of V_{T0} and V_{TE}.

A crude way to determine the threshold of a transistor is to configure it with the gate connected to the drain. The slope of the diode curve is extrapolated to crossover at the "0" current axis (Figure 5.6). This point is the V_{T0} value. Another method is to measure the voltage for two currents and use an equation to find V_{T0}.

5.5 DEPLETION TRANSISTORS

The depletion transistor characteristics are similar to those of enhanced transistors except that the depletion transistor is normally biased with $V_{GS} = 0$ (Figure 5.7). It requires negative gate-to-source voltage to turn it off. The voltage to turn the transistor off is the depletion threshold voltage. In Figure 5.7, the threshold voltage, V_{TD}, would be −3.5 V.

The voltage at which the depletion transistor is turned off is sometimes called the *pinch-off voltage*. This is a carryover from discrete transistors having two gates, which, when properly biased, would actually pinch off the channel between the source and drain.

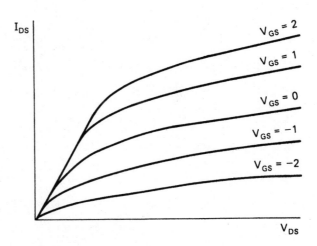

FIGURE 5.7 I_{DS} vs V_{DS} characteristics of a depletion NMOS transistor.

5.6
FAN-OUT AND
CHARGING TIME

Fan-out refers to the number of different gates that a node may drive. This is useful in determining loading when working with TTL, MOS, and discrete components. On an IC, the design engineer is concerned with the current and the total capacitance a node drives. The engineer needs to know the sizes of all the transistors and the type of interconnect, as well as the capacitance of the gate itself. All this must be considered in order to size the load and to achieve the desired rise time and propagation delay.

5.7
ASPECT RATIO

The aspect ratio is discussed on page 47 and illustrated in Figure 3.11.

5.8
SCALING OF DIE:
OPTICAL

Optical shrink is the result of photoreducing a die to a smaller size than that designed. Unless the process is changed to compensate for the dimensional changes of the transistors, the electrical characteristics may be altered. Sometimes sizing of different mask geometries may be done to compensate for the dimensional changes and thus retain some desired property.

5.9
SCALING OF
PERFORMANCE
CHARACTERIZATION

The performance of MOS devices may be extrapolated within normal sizes and performance. This allows a design engineer to create scaled charts for different electrical characteristics. Being able to do this makes designs easier. An example of a typical table made from a scaling chart is presented in Chapter 8.

5.10
FIGURING GATE
WIDTHS AND LENGTHS

Figure 5.8 shows the average gate width and length calculations. In Figure 5.8a, the first bend does not count; all others are one-third gate length. Small jogs up to 2 μm do not count as a bend in the gate width (Figure 5.8b).

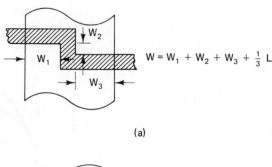

(a)

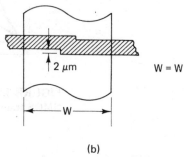

(b)

FIGURE 5.8 Method for calculating effective drawn width of an NMOS transistor.

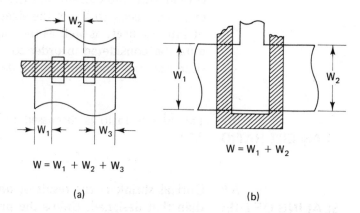

W = W₁ + W₂ + W₃   W = W₁ + W₂

(a)   (b)

FIGURE 5.9 Segmented widths and method for calculating their effective drawn width.

Gates may be broken as long as the total W/L ratio retains the same effective conductivity. Sometimes wide gates are broken to preserve the characteristics of a narrow gate. These are known as *split gates* (Figure 5.9).

**5.11
EFFECTS OF
THRESHOLD VOLTAGE
AND EXTERNAL
CONDITIONS
ON PERFORMANCE**

Table 5.1 illustrates that the same environmental conditions will result in different performances for input buffers and internal circuitry. Since input buffers are required to meet fixed input levels regardless of the process variations, their characteristics will be different from the internal circuitry, which interfaces with devices of similar characteristics. The terms *best-case speed* (or BC speed), *worst-case speed* (or WC speed), *worst-case voltage output high* (or WC V_{OH}), and

TABLE 5.1 Input Buffer and Internal
Circuitry Performance Characteristics

Characteristic	Input buffer	Internal circuitry
Worst-case speed	V_{DD} min. Temp max.	V_{DD} min. Temp max.
Best-case speed	V_{DD} max. Temp min.	V_{DD} max. Temp min.
Worst-case V_{OH}	V_{DD} min. Temp max.	V_{DD} min. Temp max.
Worst-case V_{OL}	V_{DD} max. Temp min.	V_{DD} max. Temp min.

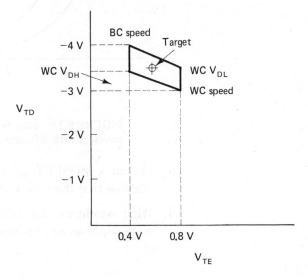

FIGURE 5.10 Relationship of enhancement transistor thresholds to depletion thresholds for NMOS transistor resulting from processing.

worst-case voltage output low (or WC V_{OL}) are derived from performance characteristics and are linked with certain combinations of transistor threshold variations (Figure 5.10). The relationship between depletion and enhancement threshold voltage is also shown in Figure 5.10.

EXERCISES

1. Using P-channel transistors, draw NAND and NOR gates for both positive and negative logic.

2. Using the load w/l propagation scale chart on page 44, determine the w/l values for 300 nanoseconds (ns) and 500 ns.

3. Fill in the current for device sizes 12/6, 10/8, 10/15, and 10/25 on the chart showing load size and current drawn on page 45.

4. Using the charts shown in Figure 5.11, determine a good ratio for an inverter; also select the "0" and "1" logic levels.

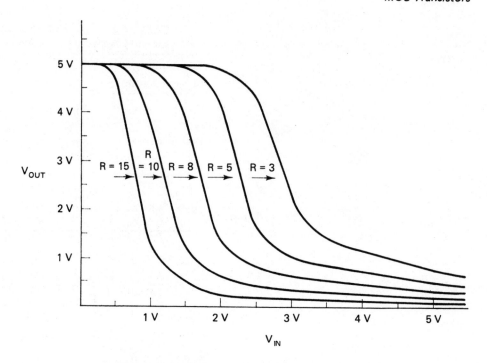

FIGURE 5.11 Transfer characteristics of simple MOS inverters with different switch-to-load ratios.

5. Sketch a MOSFET as it would appear on a composite drawing. Assume that the width and length are both 10 μm.

6. What would be the difference in the sketch of Exercise 5 if it were depleted or enhanced?

TYPICAL N-CHANNEL SILICON-GATE PROCESS

MOS processing is a technique wherein the electrical properties of a single-crystal semiconductor wafer are changed by chemical means. The changes include the creation of field-effect transistors, resistors, capacitors, and electrical interconnections. To create the desired changes in electrical properties, films are grown on the surface, ions are implanted in the surface, and chemicals are deposited on the wafer. To remove the excess film and deposited chemicals and to control the ions to be implanted, a photosensitive, etch-resistant chemical is used. A thin film of the photoresist in conjunction with masks allows the creation of a pattern. The patterns permit selective etching of the surface and the masking of ions from the wafer. The order in which wafers are treated is known as the *process flow*. Process flows vary to achieve different qualities. Flows will also vary from process line to process line to achieve the same sort of end product. Process flows are broken up into critical operations and steps, such as etching, rinsing, buffering, drying, and baking. The steps and operations are identified with the mask used at that point in the flow. An operation in this text consists of several steps.

The *5-V N-channel silicon-gate depletion-load MOS process* was developed in the early 1970s. The process is still in use today. It was the forerunner of the high-density MOS (HMOS) process, the major changes being in the techniques developed that allow the use of smaller geometric dimensions. Because of the similarities in layout rules with the original process and the modern HMOS and CMOS processes, the novice mask designer will have an easy transition learning the layout rules for newer processes. As a result, the 5-V N-channel silicon-gate depletion load MOS process was chosen for this book.

The mask numbering for the 5-V N-channel silicon-gate process allows for nine masking steps. The nine possible masks are used in seven separate operations. Each operation is unique and must be in proper sequence if the integrated circuit is to work properly. The integrated-circuit designer needs to understand the function of each mask in the process flow. It is the purpose of this chapter to illustrate in sufficient detail how each mask is used in the process flow. Thirty-four steps are illustrated.

6.1
01 MASKING OPERATION

The 01 mask is used to create the pattern that defines the field regions from the electrically active regions in an integrated circuit. The steps involved in the 01 operation begin after the wafer is cleaned. The first step is to grow a thin film of thermal oxide on the surface of the wafer. This is done by heating the wafer and subjecting it to oxygen. This layer is called *bottom oxide* (Figure 6.1). It will protect the surface of the substrate from the nitride and provide a means to control the position of implanted ions in later steps.

A thin film of nitride is deposited on top of the bottom oxide (Figure 6.2). A film of photoresist is spun on top of the nitride (Figure 6.3). The photoresist is dried and the 01 mask is aligned to the wafer. The photoresist is then exposed and developed (Figures 6.4 and 6.5).

Next, the surface of the wafer is implanted with boron ions (Figure 6.6). The boron ions in the substrate will cause the substrate to become more heavily doped with P-type characteristics. This will make the surface more difficult to invert by shifting the threshold voltage.

The next step is to etch the nitride (Figure 6.7). After the 01 mask pattern is etched into the nitride layer, the photoresist is stripped from the wafer (Figure 6.8). The next step is to grow the

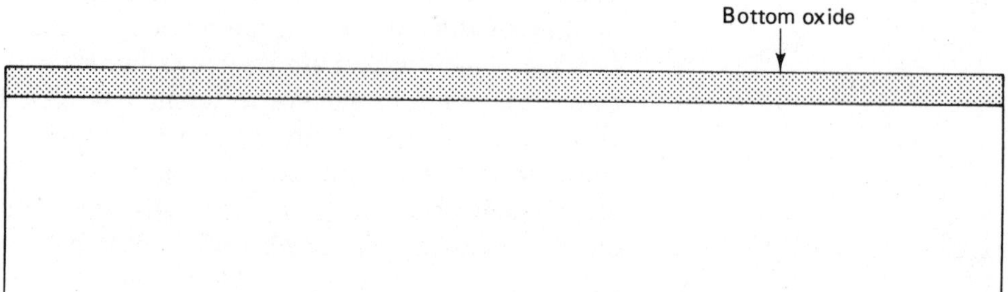

Bottom oxide

P⁻

FIGURE 6.1   Bottom oxide.

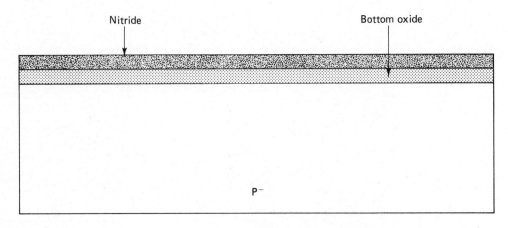

FIGURE 6.2 Nitride deposition.

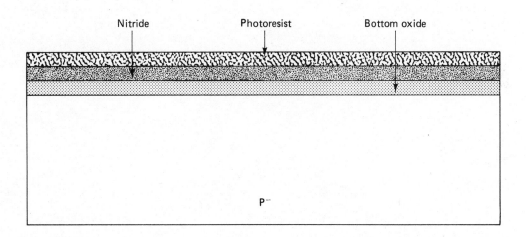

FIGURE 6.3 Photoresist spin.

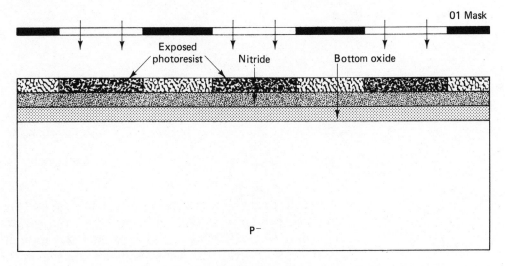

FIGURE 6.4 01 align and expose.

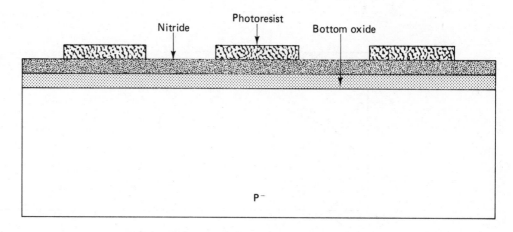

FIGURE 6.5 Develop and post-bake.

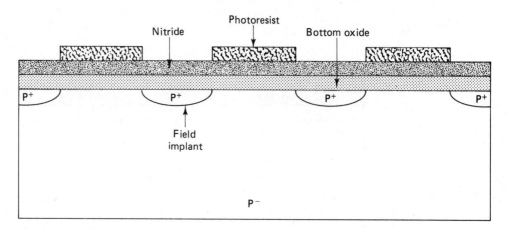

FIGURE 6.6 Boron implant.

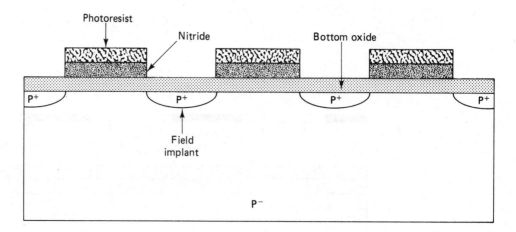

FIGURE 6.7 Nitride etch.

field oxide by exposing the wafer to oxygen (Figure 6.9). The oxygen combines with the silicon, Si, of the substrate to form silicon dioxide, SiO_2. The nitride prevents the growth of oxide in areas where it covers the wafer. Some of the silicon that had received the boron implant is converted into SiO_2. Along the edges of the nitride,

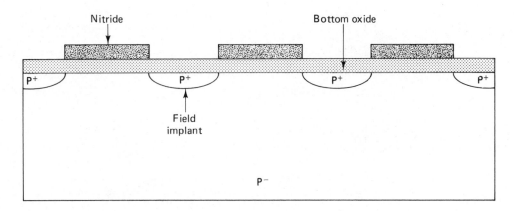

FIGURE 6.8 Strip photoresist.

Field oxidation

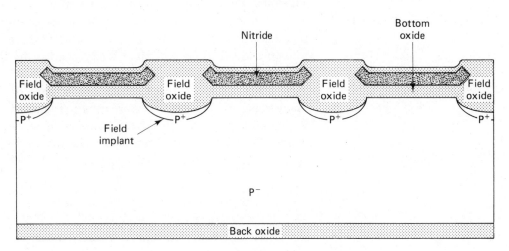

FIGURE 6.9 Field oxidation.

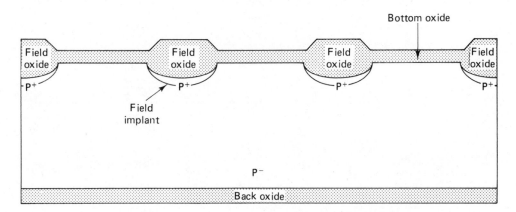

FIGURE 6.10 Top oxide and nitride strip.

the formation of the SiO_2 causes a lifting of the nitride. This lifting is referred to as "bird's beaking." SiO_2 is also formed on the back of the wafer. This layer on the back is called *back oxide*.

Next, the nitride is removed (Figure 6.10). After removal of the nitride layer, the bottom oxide is removed (Figure 6.11).

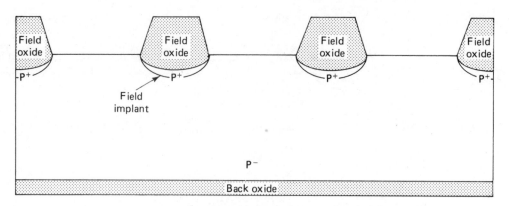

FIGURE 6.11   Bottom oxide etch.

6.2
02 MASKING OPERATION

The purpose of the 02 mask is to define the depletion transistors. This is done by permitting implantation of phosphorus ions into regions where depletion transistors will be formed. The additional ions in the substrate will electrically change the threshold voltage of those transistors.

After stripping the bottom oxide, gate oxide is grown and a blanket gate ion implant is performed (Figures 6.12 and 6.13). This implant shifts all the threshold voltages of all the transistors on the wafer. After the implant photoresist is spun onto the wafer surface,

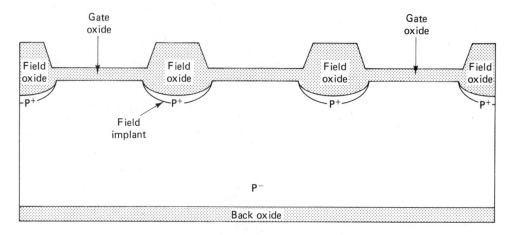

FIGURE 6.12   Gate oxidation.

FIGURE 6.13   Gate implant.

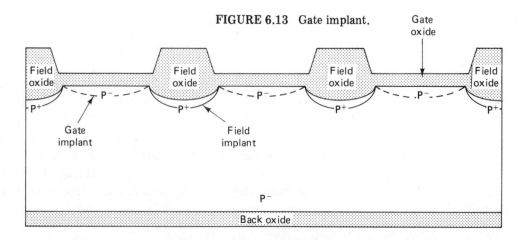

the photoresist is dried. The 02 mask is aligned to the alignment keys defined by the 01 layer and the photoresist is exposed and developed (Figure 6.14). The wafers are now exposed to the depletion implant (Figure 6.15). After implantation of the depletion transistor regions, the photoresist is stripped (Figure 6.16).

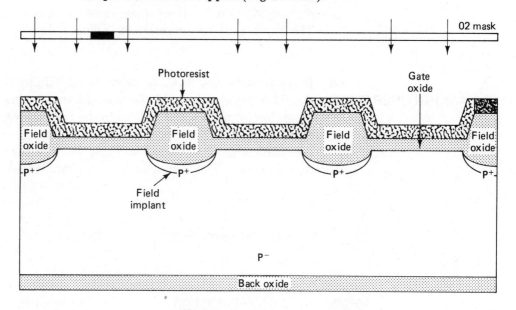

FIGURE 6.14 02 align and expose.

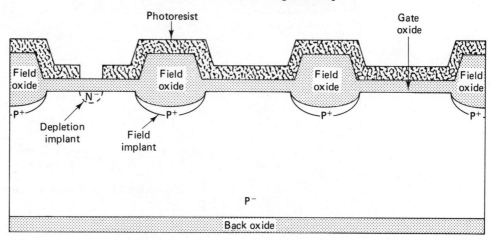

FIGURE 6.15 Depletion implant.

FIGURE 6.16 Strip PR.

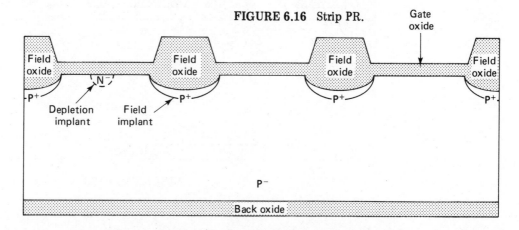

6.3
03 MASKING OPERATION

The 03 mask creates the buried contact by defining areas where gate oxide will be removed. After stripping of the photoresist with the 02 pattern, more photoresist is spun on and dried. The 03 mask is aligned to alignment keys defined by the 01 mask (Figure 6.17). The photoresist is exposed and developed. The wafer is placed into an etching solution and the exposed gate oxide is stripped (Figure 6.18). The photoresist is then removed (Figure 6.19).

6.4
04 MASKING OPERATION

If required by the flow, an 04 mask containing the same pattern as the 03 mask may be used. In this case the etching in the 03 masking operation is omitted and is done in the 04 masking operation. The 04 mask permits an additional photoresist layer to be used, eliminating the possibilities of etchant working its way through a single layer of photoresist and removing gate oxide. Where the gate oxide is

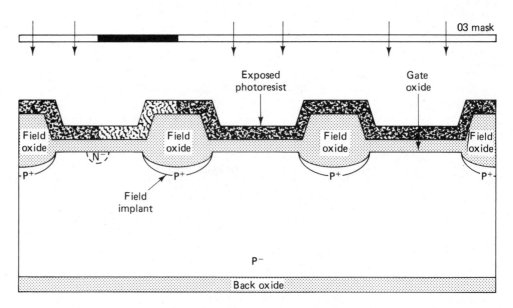

FIGURE 6.17 03 align and expose.

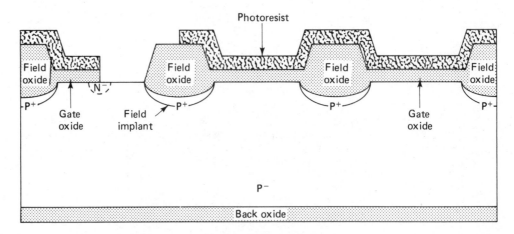

FIGURE 6.18 03 gate etch.

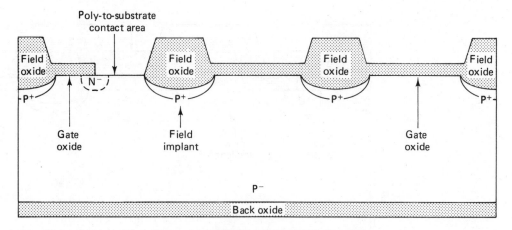

FIGURE 6.19 Photoresist strip.

removed, the polycrystalline silicon film deposited on the surface will come into contact with the substrate. Except for this paragraph, the 04 masking operation is not described in this text.

6.5 05 MASKING OPERATION

A polycrystalline silicon* film is deposited on the surface of the wafer (Figure 6.20), and the surface of the film is oxidized (Figure 6.21). The 05 mask defines the gates of the transistors and capacitors and the poly interconnect lines. After poly oxidation, the photoresist is spun onto the surface of the wafer. The 05 mask is aligned to alignment keys defined by the 01 mask and the photoresist is exposed and developed (Figure 6.22). The 05 pattern is etched into the poly oxide (Figure 6.23). The photoresist is then removed (Figure 6.24). Using the poly oxide as a mask, the 05 pattern is etched into the poly (Figure 6.25). When the excess poly is removed, the substrate for the sources and drains and N+ interconnect will be exposed (Figure 6.26).

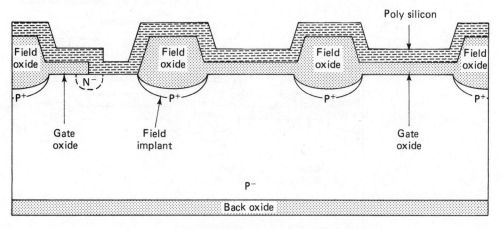

FIGURE 6.20 Polycrystalline silicon.

*In the industry, the term "poly" is commonly used for "polycrystalline silicon," and we will use that term extensively in the following text.

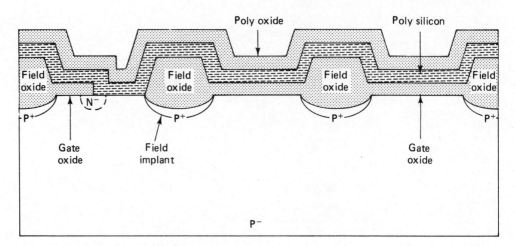

FIGURE 6.21   Polycrystalline silicon oxidation.

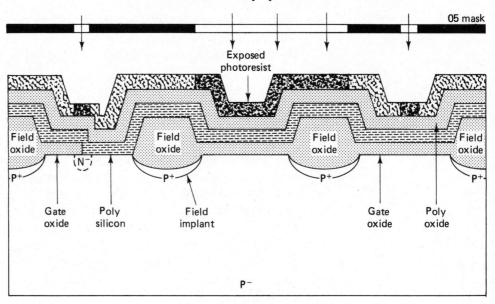

FIGURE 6.22   05 align and expose.

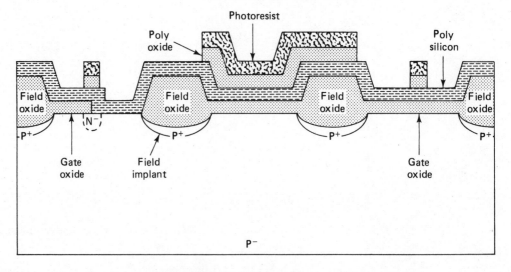

FIGURE 6.23   Polycrystalline silicon oxide etch.

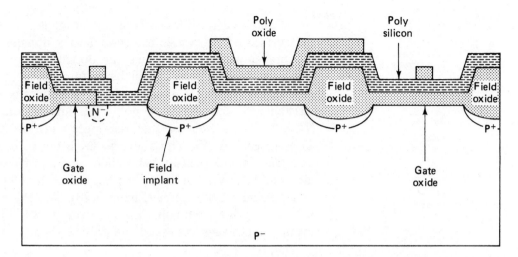

FIGURE 6.24 Photoresist strip.

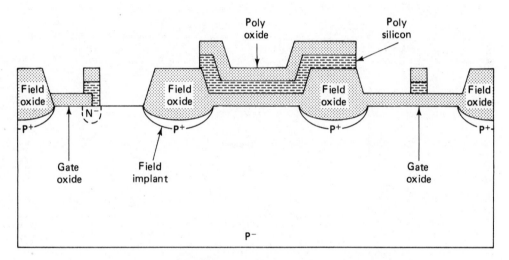

FIGURE 6.25 Polycrystalline silicon etch.

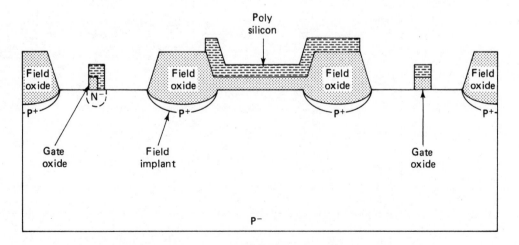

FIGURE 6.26 Gate oxide etch.

Following the 05 masking operation is a predeposition where the surface of the wafer is exposed to a dopant. The objective of the predeposition is to place the dopant in contact with the raw silicon surface, where it will be diffused into the silicon. (In this process the dopant is phosphorus.) The dopant is used to reduce the resistance of the deposited polycrystalline silicon and to form the N⁺ source and drain of the transistors and the N⁺ interconnect regions (Figure 6.27). The wafer is subjected to heat which "drives" or diffuses the dopant into the silicon and also forms a film of silicon dioxide. After the "drive-in," a phosphorus-doped (Phossil) glass film is deposited on the wafer (Figure 6.28). The thermally grown oxide and the deposited Phossil glass form the field oxide and provide the insulation between the metal and poly layers and the N⁺ regions.

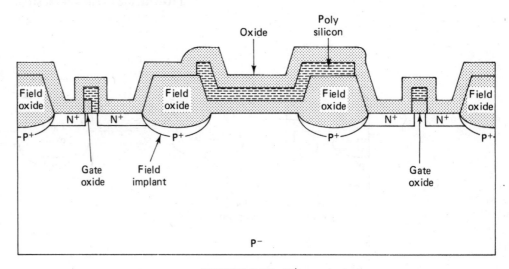

FIGURE 6.27 N⁺ source–drain.

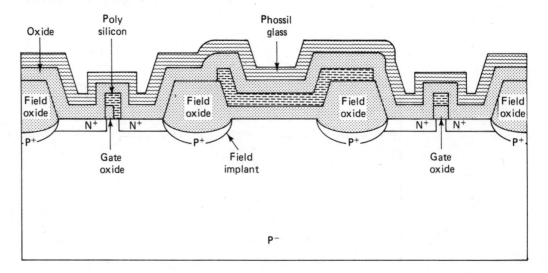

FIGURE 6.28 Glass deposition.

6.6
06 MASKING OPERATION

The 06 mask defines the preohmic contact cuts. These contacts are the result of metal passing through the insulating layers to the poly or N$^+$ regions. After the Phossil glass deposition, photoresist is spun onto the surface of the wafer and dried. The 06 mask is aligned to the alignment keys defined by the 05 mask and the photoresist is exposed and developed (Figure 6.29). The wafer is exposed to an etchant. The 06 contact holes are then etched (Figure 6.30).

6.7
07 MASKING OPERATION

Occasionally, an 07 mask is used to taper the holes created by the 06 mask. The pattern is the same on the 07 as on the 06 except for the size.

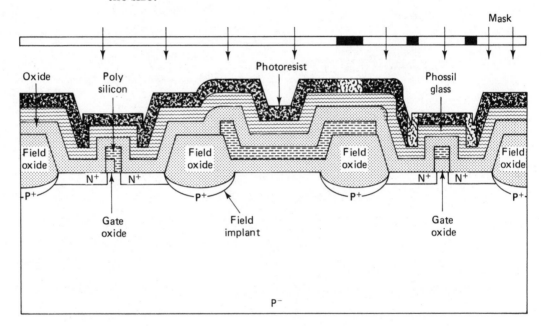

FIGURE 6.29   06 align and expose.

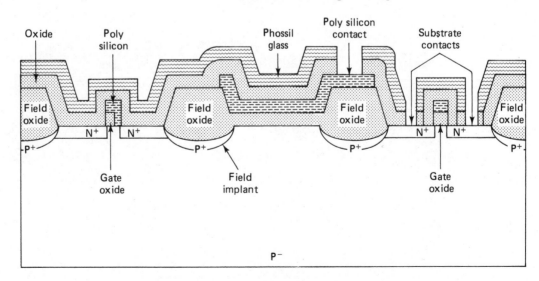

FIGURE 6.30   Preohmic contact cut and photoresist strip.

6.8
08 MASKING OPERATION

The 08 mask defines the interconnect, power buses, and the pads for the integrated circuit. Once the 06 contact holes have been cut, a metal film is deposited on the surface of the wafer (Figure 6.31). Photoresist is again deposited on the surface of the wafer. The photoresist is then dried and the 08 mask aligned to the alignment keys defined by the 06 mask (Figure 6.32). Next, the photoresist is exposed and developed, and the metal pattern is etched (Figure 6.33). The metal photoresist is then removed. Passivation is deposited on the metal pattern (Figure 6.34).

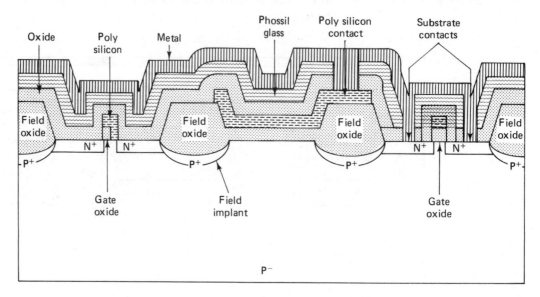

FIGURE 6.31   Metal.

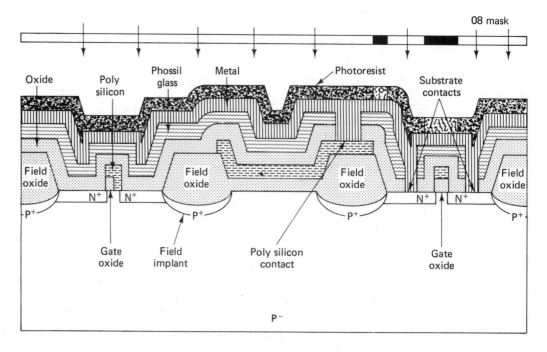

FIGURE 6.32   08 align and expose.

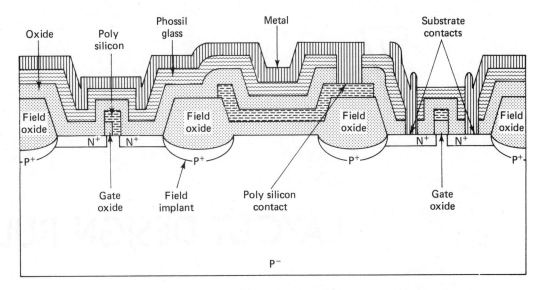

FIGURE 6.33 Etch and photoresist strip.

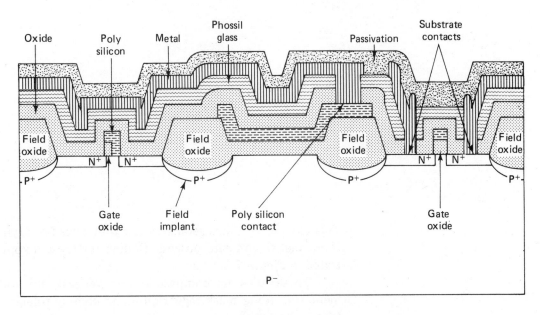

FIGURE 6.34 Final passivation.

6.9
09 MASKING OPERATION

In this step, another film of photoresist is spun onto the surface of the wafer and dried. The 09 mask, which defines the openings for the pads, is aligned to the alignment keys defined by the 08 mask. The photoresist is exposed, developed, and etched. At this point the wafers are ready for back lap and back plating. After plating, the wafers are probed and sent to the assembly area for packaging.

7

LAYOUT DESIGN RULES

This chapter contains geometrical layout rules for a typical MOSFET N-channel silicon gate process. Coding for these layout rules is illustrated in Figure 7.1.

These rules are compatible with vertical and horizontal alignment (including mask registration and etching tolerances of $\pm 2\mu$m). It is to be noted that simultaneous worst-case alignment in both the X and Y directions will result in a 2.8μm-misalignment in a diagonal direction. This effect is represented in Figure 7.2. To compensate for this 40% increase in alignment tolerance, a second dimension column for diagonal alignment is noted in the rules and must be strictly obeyed.

Dimensional tolerances of masks are expected to be less than $\pm 1\mu$m. Finished line-width dimensions will be skewed during processing, as indicated in Table 7.1.

Following are the layout design rules.

Layer	Mask No.	Symbol	Description
Active layer	01		Solid outline
Depletion mask	02		Outline of short dashes
Poly to N$^+$	03	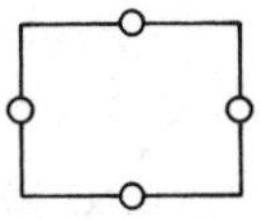	Solid outline with small circles
Oversize 03	04		Not normally drawn
Polycrystallized silicon (gate)	05	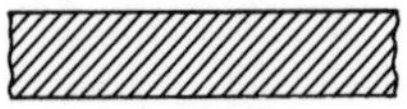	Solid outline with hatched area
Contact holes (preohmic)	06		Solid outline with cross
Oversize 06	07		Not normally drawn
Metal	08	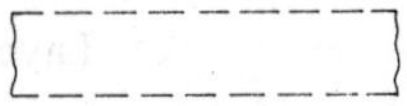	Outline of long dashes
Passivation	09		Outline consisting of long and short lines

FIGURE 7.1 Coding for layout rules.

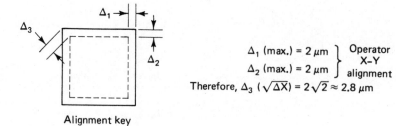

Alignment key

Δ_1 (max.) $= 2\ \mu m$ } Operator X–Y alignment
Δ_2 (max.) $= 2\ \mu m$
Therefore, $\Delta_3\ (\sqrt{\Delta X}) = 2\sqrt{2} \approx 2.8\ \mu m$

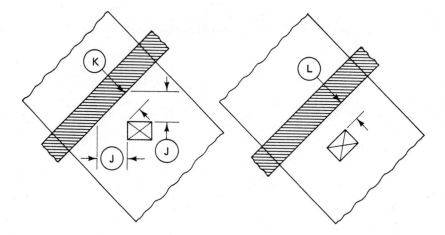

Examples of diagonal alignment tolerances

FIGURE 7.2 The example in this figure refers to rule 7.6.3.2; dimensions K and L above correspond to dimensions A in Figure 7.21.

TABLE 7.1 Masking Operations

Layer	Mask number	Density (for negative resist)	Align to:	Nominal etched dimension
Active area	01	Dark field	Wafer	−0.8*
Depletion mask	02	Light field	01	−0.75
Poly-N^+ contact	03	Light field	01	−0.75
Oversize 03	04	Light field	01	—
Gate	05	Dark field	01	0.0
Preohmic	06	Light field	05	+0.8
Oversize preohmic	07	Light field	06	—
Metal	08	Dark field	06	−2.0
Passivation	09	Light field	08	+1.0

*Includes size adjustment due to field oxide encroachment.

	Figure reference	Normal dimensions (μm)	Diagonal dimensions
7.1 ACTIVE AREA (01 MASK)			
7.1.1 Minimum width	7.3-A	6	6
7.1.1.1 Minimum width for critical-valued resistors	7.3-B	8	8
7.1.1.2 Minimum width for critical active devices	7.3-C	8	8
7.1.2 Minimum space	7.3-D	6	6
7.2 DEPLETION LOAD (02 MASK)			
7.2.1 Minimum overlap of depletion device			
7.2.1.1 If 02 edge is on diffusion	7.4-A	5	7
7.2.1.2 If 02 edge is on field oxide	7.4-B	3	4
7.2.2 Minimum space from enhancement device			
7.2.2.1 If 02 edge is on diffusion	7.4-C	5	7
7.2.2.2 If 02 edge is on field oxide	7.4-D	4	4
7.3 POLY-N⁺ CONTACT (03 MASK)			
7.3.1 Minimum dimension on 03 contact mask			
7.3.1.1 In direction of N⁺	7.5-A	10	13
7.3.1.2 In other directions	7.5-B	10	13
7.3.2 Minimum dimensions of contact region	7.5-C, D	6 $\times$ 6	7 $\times$ 7
7.3.3 Minimum extension of 03 beyond contact area			
7.3.3.1 In direction of N⁺	7.5-F	5	7
7.3.3.2 In all other directions	7.5-E	2	3
7.3.4 Minimum space from active gate			
7.3.4.1 If 03 is on field oxide	7.6-A	5	6
7.3.4.2 If 03 edge is on diffusion	7.6-B	5	6
7.3.5 Design layout caution *Capacitance:* Thinning of the field oxide occurs when the 03 cut is on field oxide. If the capacitance to the substrate of adjacent poly is important, it should be placed away from the 03 cut.			
7.3.5.1 Spacing of 03 from 05 on field for minimal increase in poly-to-substrate capacitance	7.6-C	5	6

		Figure reference	Normal dimensions (μm)	Diagonal dimensions
7.3.6	Minimum space between 06 contact region and poly–N$^+$ contact (03) region			
	7.3.6.1 Contact to poly	7.7-A	5	7
	7.3.6.2 Contact to N$^+$	7.7-B	5	7
7.3.7	03 Contact to large N$^+$ region when N$^+$ extends beyond the contact area in all three directions by at least 3.0 μm (4 μm diag.), the minimum extension of 03 beyond the contact area is:	(7.8-A)	(3)	(4)
		7.8-B	2	2

7.4 OVERSIZE POLY-N$^+$ CONTACT
Pattern is to be machine-generated by increasing the 03 by 6 μm (3 μm on each side)

		Figure reference	Normal dimensions (μm)	Diagonal dimensions
	7.4.0.1 Minimum dimensions on 04 mask		16 $\times$ 19	

Note: Oversizing this mask may cause separate contacts to a common area to merge into one large contact.

7.5 POLYCRYSTALLINE SILICON GATE (05 MASK)

		Figure reference	Normal dimensions (μm)	Diagonal dimensions
7.5.1	Minimum width	7.9-A	6	6
7.5.2	Minimum space	7.9-B	6	6
7.5.3	Minimum gate length	7.9-C	6	6
7.5.4	Minimum poly overlap of field oxide	7.10-A	5	6
7.5.5	Minimum space poly on field to adjacent N$^+$	7.11-A	2	3
	7.5.5.1 Adherence to the rule above will ensure that overlap between poly and N$^+$ does not occur			
	7.5.5.2 Packing density can be increased by reducing A in Rule 7.5.6.2 with the following electrical penalties: (a) Increased poly-to-substrate or poly-to-N$^+$ capacitance (b) Increased resistance			
7.5.6	Minimum space for poly on diffusion to field (01) edge			
	7.5.6.1 If poly runs parallel to 01 edge for less than 50 μm	7.12-A	6	8

	Figure reference	Normal dimensions (μm)	Diagonal dimensions
7.5.6.2 If poly runs parallel to 01 edge for greater than 50 μm	7.12	7	9

Warning: Series resistance in source-drain areas of device significantly reduces the effective W/L ratio of the device.

Note: In the case of very wide devices, the source and drain regions should be strapped with metal as shown in Figure 7.13. The 06 contacts may be joined.

7.6 PREOHMIC CONTACTS (06 MASK)

	Figure reference	Normal dimensions (μm)	Diagonal dimensions
7.6.1 Size			
7.6.1.1 Metal (08)-to-N$^+$ (01) contact	7.14-A$_1$	5×7	5×7
	7.14-A$_2$	6×6	6×6
7.6.1.2 Metal-(08)-to-N$^+$ (01) contact minimum spacing; multiple contacts are required in lieu of large contact to a common area	7.15-A	5	5
7.6.1.3 Metal-(08)-to-poly (05) contact	7.16-B$_1$	5×7	5×7
	7.16-B$_2$	6×6	6×6
7.6.1.4 Metal-(08)-to-poly (05) contact minimum spacing; multiple contacts are to be used in lieu of a large contact to a common area	7.17-A	5	5
7.6.2 Minimum space from N$^+$ (01) edge			
7.6.2.1 Contact to N$^+$ (01)	7.18-A	3	4
7.6.2.2 Contact to poly (05)	7.19-A	5	6
7.6.3 Minimum space from poly edge			
7.6.3.1 Contact to poly			
7.6.3.1.1 In direction of metal line	7.20-A	4	6
7.6.3.1.2 In other directions	7.20-B	3	4
7.6.3.2 Contact to N$^+$ in the direction of the transistor or capacitor	7.21-A	5	7

7.7 OVERSIZE PREOHMIC (07 MASK)

Oversize 07 mask is to be machine-generated by increasing the dimensions a total of 8 μm (4 μm on each side)

	Figure reference	Normal dimensions (μm)	Diagonal dimensions
Size (for 5×7 preohmic)		13×15	13×15
Size (for 6×6 preohmic)		14×14	14×14

	Figure reference	Normal dimensions (μm)	Diagonal dimensions

Note: Oversizing this mask will cause separate contacts to a common area to merge into one large contact.

7.8 METAL (08 MASK)

		Figure reference	Normal dimensions (μm)	Diagonal dimensions
7.8.1	Minimum line width	7.22-A	8	8
7.8.2	Minimum space	7.22-B, C	7	7
7.8.3	Minimum overlap of contact opening			
7.8.3.1	Contact to poly (05)	7.23-A	2	3
7.8.3.2	Contact to N⁺ (01)	7.23-B	2	3
7.8.4	Bonding pads			
7.8.4.1	Minimum size	7.24-A	138 × 138	
7.8.4.2	Minimum pad-to-pad space	7.24-B	138	
7.8.4.3	Minimum space to any component	7.24-C	32	
7.8.4.4	Minimum space to inside of scribe line from outer edge of passivation (09)	7.24-D	54	
7.8.4.5	Minimum space to any process key	7.25	12	

7.9 PASSIVATION (09 MASK)

		Figure reference	Normal dimensions (μm)	Diagonal dimensions
7.9.1	Passivation mask to be inside bonding pad	7.24-E	5	7
7.9.2	Minimum dimensions (e.g., internal probe pads)		25	25
7.9.3	Minimum space to any component	7.24-F	37	

7.10 CHIP PERIPHERY REQUIREMENTS

		Figure reference	Normal dimensions (μm)	Diagonal dimensions
7.10.1	Alignment keys (shown in Figure 7.25) are to be placed near the center of the long edge of chip ≥ 10.0 μm from scribe or active circuit elements	7.25		
7.10.2	The appropriate test device should be placed on the periphery of the chip, maintaining 10-μm spacing to other elements	7.26		
7.10.3	Input pads (standard input protect network)	7.27		
7.10.4	Poly ground ring must be included on chip periphery			
7.10.4.1	Distance from metal	7.28-A	35	
7.10.4.2	Width of poly ring	7.28-B	14	
7.10.5	Scribe grid to be 3 mils wide	7.28-C	75	

		Figure reference	Normal dimensions (μm)	Diagonal dimensions
7.10.6	Distance of other layers from metal pad			
7.10.6.1	Active area (01)	7.24-C	42	
7.10.6.2	Depletion mask (02)	7.24-C	42	
7.10.6.3	Poly-N$^+$ contact (03)	7.24-C	42	
7.10.6.4	Contact holes (06)	7.24-C	42	
7.10.6.5	Passivation (09)	7.24-C	54	
7.10.7	Logo must be placed inside scribe line; may be on 01 or 08 layer	7.29		

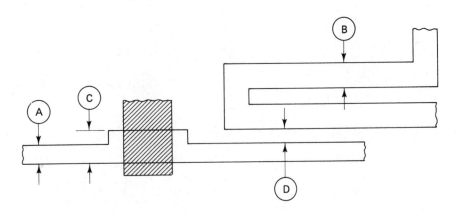

FIGURE 7.3   Active area (01 mask).

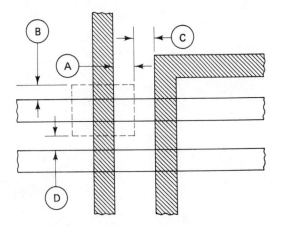

FIGURE 7.4   Depletion load (02 mask).

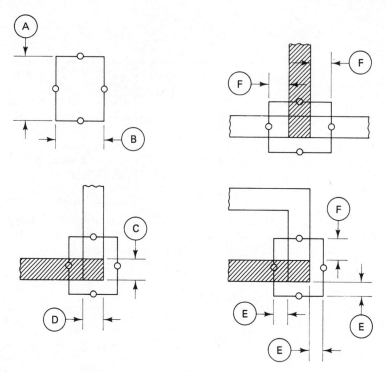

FIGURE 7.5 Poly-N⁺ contact (03 mask).

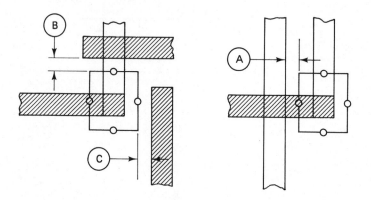

FIGURE 7.6 Poly-N⁺ contact minimum critical spacing.

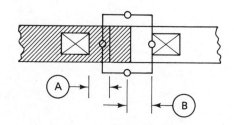

FIGURE 7.7 Poly-N⁺ contact and 06 contact.

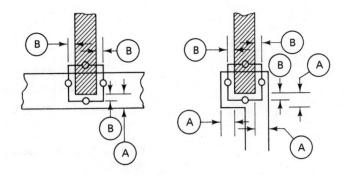

FIGURE 7.8 03 contact to N$^+$ regions.

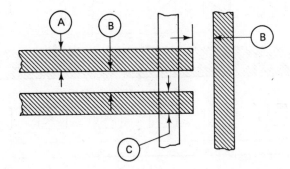

FIGURE 7.9 Polycrystalline silicon gate (05 mask).

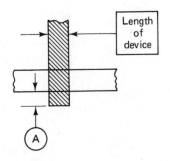

FIGURE 7.10 Minimum poly overlap.

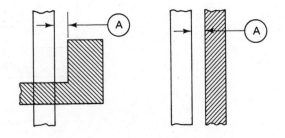

FIGURE 7.11 Minimum space of poly to N$^+$.

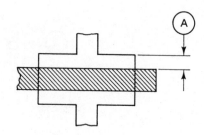

FIGURE 7.12 N⁺ transistor source or drain overlap of poly.

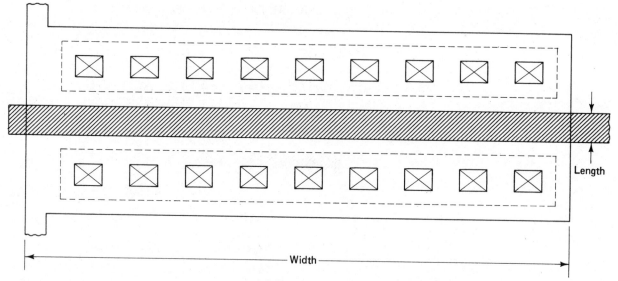

FIGURE 7.13 Transistor source and drain resistance may be reduced by using 06 contacts and metal lines.

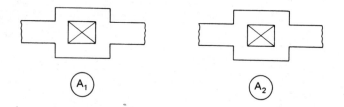

FIGURE 7.14 Preohmic (06 contact) 06 mask.

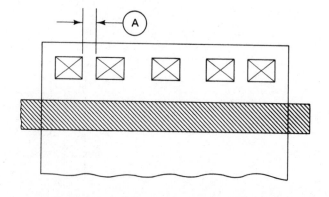

FIGURE 7.15 Metal-(08)-to-N⁺ (01) contact.

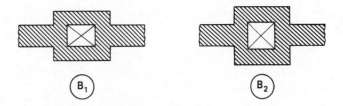

FIGURE 7.16 Metal-(08)-to-poly (05) contact.

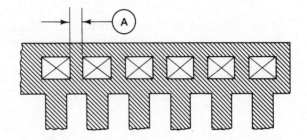

FIGURE 7.17 Minimum 06-to-06 contact spacing.

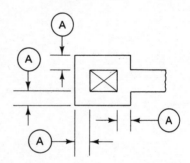

FIGURE 7.18 Minimum N$^+$ (01) overlap of (06) contact.

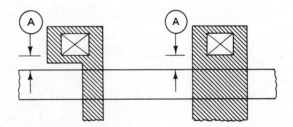

FIGURE 7.19 Minimum space of (06) contact over field region to transistor gate.

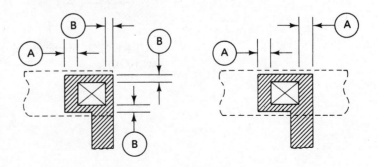

FIGURE 7.20 Minimum poly (05) overlap of (06) contact.

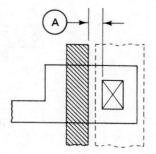

FIGURE 7.21 Minimum space of (06) contact over N⁺ (01) region to transistor gate.

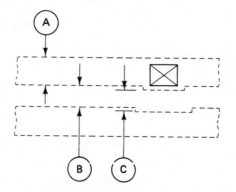

FIGURE 7.22 Metal (08 mask).

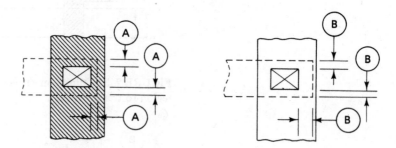

FIGURE 7.23 Minimum metal (08) of preohmic contact (06).

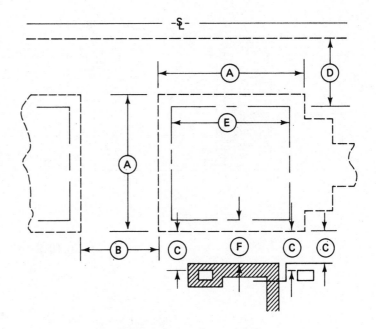

FIGURE 7.24 Bonding pads.

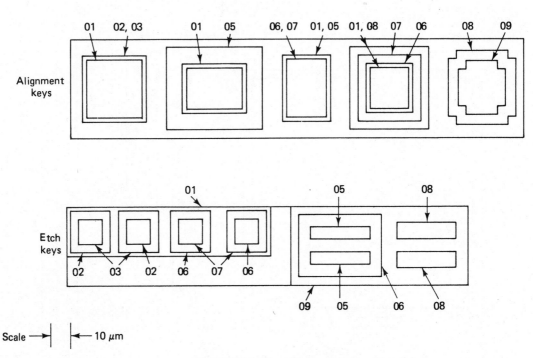

FIGURE 7.25 Alignment keys and etch keys.

Test site

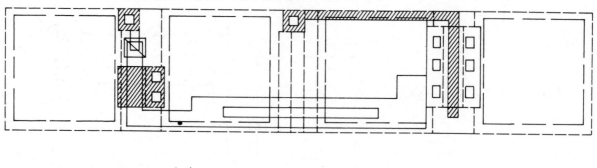

Scale ──▶| |◀── 10 µm

FIGURE 7.26   Test site.

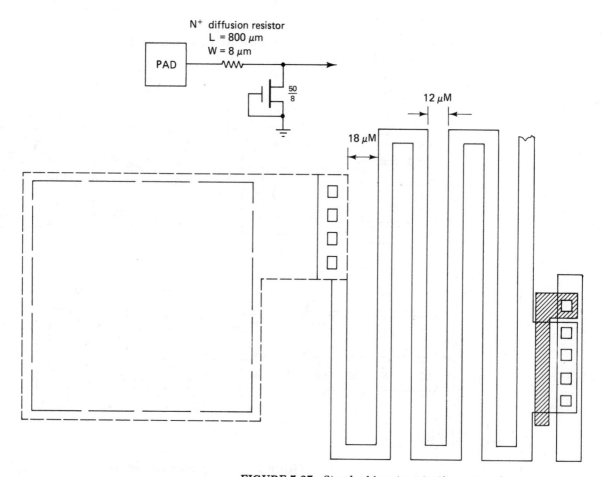

N^+ diffusion resistor
L = 800 µm
W = 8 µm

PAD

$\frac{50}{8}$

12 µM

18 µM

FIGURE 7.27   Standard input protection network.

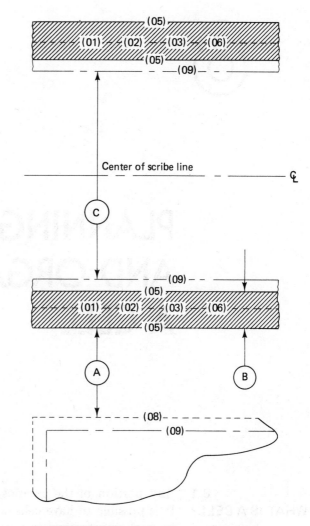

FIGURE 7.28 Scribe line requirements.

LOGO

Either 01 or 08 layer

Scale: →| |← 10 μm

FIGURE 7.29 Logo.

PLANNING AND ORGANIZING A CELL

8.1
WHAT IS A CELL?

Any portion of the composite drawing may be designated as a cell. It is possible to have cells within cells. The composite drawing may be broken into smaller portions to facilitate drawing or digitizing. A cell may be as simple as containing a single geometric shape or may contain the whole composite drawing. In practice, a designer will define repetitive shapes as a cell to reduce drawing time. Frequently, the unchanged portion of similar structures is made into a cell. The changed portions each may become cells. The use of Calma digitizing systems permits the nesting of cells within cells. This system of nesting allows the designer great freedom in creating drawings.

The elements associated with a cell are (1) the origin; (2) the height and width, or area; (3) the orientation; and (4) the components. Each cell will be given a name (Figure 8.1). If the cell is placed on a drawing, its name, orientation, and area must be defined. To define the cell boundaries, a blue rectangle is placed around the cell. All the geometrics associated with the cell should be within the blue rectangle (occasionally, exceptions may be made). Whenever the cell is placed on a composite drawing, the symbols indicated in Figure 8.2 will be shown.

Figure 8.3a illustrates the different cell orientations and views that are used in this text. The cell may also be mirrored (Figure 8.3b). Note that the rotations also go clockwise (Figure 8.3c).

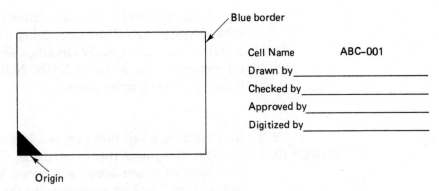

FIGURE 8.1 Requirements for cell designation showing origin and label.

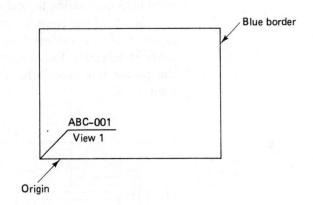

FIGURE 8.2 A cell placement showing origin and view.

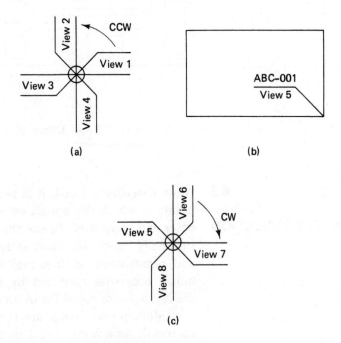

FIGURE 8.3 Cell orientations.

Each digitizing system has different schemes for orientation of cells. When planning and organizing a cell, always consider the elements of the cell. In many circuits, cells may contain clusters of logical elements such as the NANDs, NORs, and INVs (inverters) that make up some complex gates.

8.2 POWER BUS

In developing a cell that contains logical elements, it is necessary to consider the power that is available. Cells influence the choice and placement of power buses, and power buses influence the choice of cells and the topology to implement the logic.

The usual arrangement of the transistors for the gates is such that the loads will be near the V_{DD} bus and the switches will be near the V_{SS} bus. This is not a hard-and-fast rule; however, it is the way most cells containing logical elements are organized.

Much of the circuitry may be built beneath the power bus. Generally, this is considered to be good practice, as it utilizes the area more efficiently. Two chip plans showing how the organization of the power bus may influence the design of a cell are illustrated in Figure 8.4.

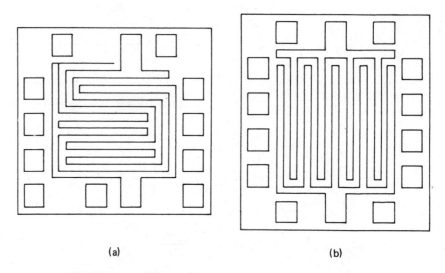

(a)   (b)

FIGURE 8.4 Effect of cell placements upon power bus and die shape.

8.3 SIGNAL INTERCONNECTION

When developing a cell, it is very important to consider the physical points at which the signals enter and leave the cell. Where the signals enter and leave may dictate the placement of the switches and loads. Internally, in a cell, most interconnect lines are short. The source and termination of lines may exist on different layers. Internally, unless otherwise specified by the design engineer, a designer may use N^+, poly, or metal for interconnection.

Most power buses are run in parallel, and they are normally on metal. As a result, most metal interconnect lines are run in parallel to the metal power buses. Other interconnect lines may be run either in parallel or perpendicular to the power buses.

8.4 AREA

When planning a chip, certain areas are allocated to portions of the logic. Based on experience and cell drawings, it is possible to estimate the area required for different logic gates. An area of 10,000 μm^2 might be allowed for a NOR gate.

One of the biggest area wasters on almost all chips is interconnect, which is necessary to implement the logic. When planning a cell, it is best to keep the interconnect area to a minimum.

8.5 BOUNDARIES

Some of the biggest problems encountered with cells by designers are at the boundaries, which is where cells interface with other cells. Placement of signal and bus lines to ensure that they meet and connect properly can become confusing. Take special care when defining cell boundaries and the associated interconnect.

8.6 CHIP PLAN

The chip plan is a plan of how the chip or die will be organized (Figure 8.5). It is very important to develop a good chip plan. A logic cell placed in the wrong place on the plan can cause the chip area to

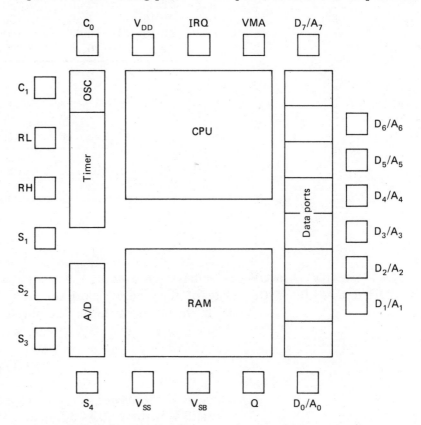

FIGURE 8.5 An example of an initial chip plan showing placement of logic blocks.

increase by several hundred micrometers. How a chip is to interface with other logic frequently influences the size and aspect ratio of a cell.

**8.7
SKETCHES**
Most designers make schematics from the logic diagrams. Not only will the sketches help in developing a cell plan, but they allow the designer to visualize the signal and bus interconnect and interface. To properly proportion the sketch, it is necessary to know the transistor sizes. The logic diagram, when properly sized by the design engineer, will show a gate size beside each logic gate or element. The transistor sizes for the gate may be found in a table provided by the design engineer. Table 8.1 illustrates the transistor sizes that will be used for this text. Table 8.2 contains the sizes for the coupler transistors.

TABLE 8.1   Device-Size Table for Logic Gates

Size	Load, W/L	NOR, NOT switch, W/L	NAND switch, W/L	Coupler-driven switch, W/L	Loading capacitance, pf
A	10/25	25/6	50/6	37/6	1/4
B	10/15	40/6	80/6	60/6	1/2
C	10/10	60/6	120/6	90/6	3/4
D	10/6	100/6	200/6	150/6	1

TABLE 8.2   Device-Size Table for Couplers

Size	W/L
T	10/8

**8.8
REDUCTION OF
NODAL CAPACITANCE**
Reduction of nodal capacitance is the name of the game: the less capacitance on a node, the less power required to change the logic state of the node. Significant reduction of nodal capacitance will result in improvement in the speed power product for the part. Dynamic power is described by the equation

$$P = CV^2 f$$

where P is the power, C is the capacitance to be charged and discharged, V is the voltage difference, and f is the frequency of transitions. Frequently, a reduction in capacitance also means a reduction in area and size.

PROJECT: DESIGN OF CELLS

9.1 PURPOSE

The purpose of this project is to help the student develop the necessary skill to convert simple logic functions into transistors. The student will design an inverter, an input buffer, a two-input NOR gate, a four-input NOR gate, and a three-input NAND gate. The difference between the simple logic functions and the complex functions is the amount of logic implemented. Details for these cells will be found in the logic diagrams or cell sheets from the design project in Appendix A, which shows the details of complex logic elements.

9.2 METHODOLOGY

First, the logic in each cell will be translated to schematics. Once the logic has been reduced to a schematic, details of the interconnect and the power bus need to be considered before further steps are taken. The direction from which the input and output signal lines must come and go will have a major impact on placement of the transistors that make up the logical elements. The position of the bus line may force certain configurations. A good reference to use prior to designing the circuit is the chip plan. The chip plan will show the specific logic elements and where the designer plans to place them.

For this project, certain assumptions will be made regarding the signal routing, and power busing. The inputs will be from the left and the outputs will be on the right. The V_{DD} bus will appear at the top. The V_{SS} bus will appear below.

When developing a chip plan, the designer will ensure that the power buses (V_{SS}, V_{DD}) reach all logical functions. Often the designer will take the option of routing the metal buses primarily in one direction and signal lines in poly in a direction perpendicular to the metal.

9.3
EXAMPLE

In this example, the power buses, V_{SS} and V_{DD}, will run from left to right. The V_{DD} bus will be at the top of the diagram and 10 μm wide. The V_{SS} bus will be 50 μm below V_{DD} and 38 μm wide. The inputs will be on poly and will enter from the left. The output will exit to the right.

This example consists of an inverter driving a NOR gate (Figure 9.1a). The inverter and NOR gate appear schematically as shown in Figure 9.1b. Many designers make a sketch of the circuit to work up ideas for possible arrangement (Figure 9.2a). Others use blocks or a stick figure. Next, the actual circuit is sketched (Figure 9.2b). Finally, the sketch is colored, checked, and prepared as a cell.

In drawing the gates, minimum N^+ and poly are used on the signal nodes. Larger amounts of N^+ are used for the V_{SS} and V_{DD} signals to reduce the voltage drops. Although a single 06 contact is sufficient to complete the circuit for this cell, additional contacts to the power buses are used to reduce the resistance.

Sometimes a circuit may be redrawn repeatedly to make it fit into a certain specific area, to mate with the bus and signal requirements, or to meet certain critical loading requirements. Since no chip

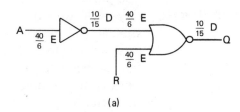

(a)

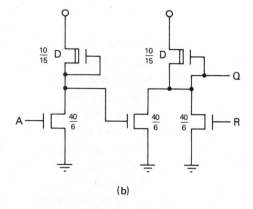

(b)

FIGURE 9.1 (a) Inverter driving a NOR gate shown logically, (b) and schematically.

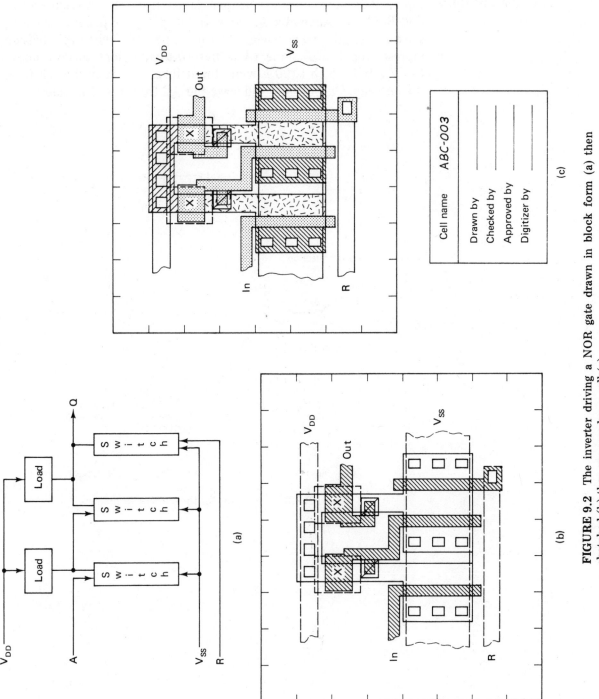

FIGURE 9.2 The inverter driving a NOR gate drawn in block form (a) then sketched (b) then prepared as a cell (c).

117

plan has yet been developed in this project, many of the cells designed for this part of the project may have to be redesigned.

CELL EXERCISES   Figure 9.3 shows several examples taken from the "Design Project Cell Book" in Appendix A. Draw sketches on grid paper of the cells shown. Use grid paper having 20 squares per inch, with each division representing 2 μ. Assume a V_{DD} bus 10 μ wide, located 25 μ above the V_{SS} bus which is 20 μ wide. Input lines will enter the cell from the left side. Output lines will leave the cell from the right side.

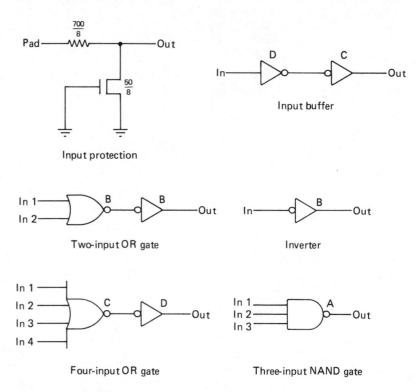

FIGURE 9.3 Other cells for the student to prepare.

10

SIZING POWER, CLOCK, AND SIGNAL BUSES

The size of the buses on a die can be critical. Power buses must be sized for current density and for voltage drops. Clock and signal buses must be sized for minimum time constants.

**10.1
PAD CONSTRAINTS**

Input signals to a die are applied to the pads via bonding wires. The signal passes through an N^+ resistor. The signal is then applied to the gate of an input transistor via a protection device. By the time the signal reaches the input transistor, it has been delayed by the time constant of the input protection scheme. With signals such as clocks, where the delay of the signal is undesirable, all or part of the protection resistor may be removed.

Power buses do not have protection resistors because the drop across the protection resistor is prohibitive.

**10.2
IR DROPS**

Power buses must be sized to keep the voltage drops due to resistance and current within limits. Therefore, V_{DD} and V_{SS} must be routed on metal. Any tunneling must be done with the design engineer's consent. Short tunnels to drains and sources are encouraged to reduce the die area. The V_{SS} bus is especially sensitive. Normally, this bus is about twice as wide as the V_{DD} bus. Wide N^+ tunnels to sources of switches are used to reduce the voltage drops to a minimum. Voltage drops at switch sources will cause changes in the threshold voltage of the switch, which is undesirable.

As a rule of thumb, the maximum voltage drop acceptable on any point of a V_{SS} line is 0.2 V. On V_{DD} lines it may be up to 0.5 V on internal logic. On output buffers it should not be more than 0.2 V.

10.3 FAN-OUT

The design engineer determines the size of the load transistor based on the fan-out of the signal and some value of capacitance based on estimates. To estimate the loading, the chip plan is used to determine the length of routing and the number of tunnels.

It is important for the design engineer to show all critical signals on the chip plan. The design engineer will designate the critical paths on the logic diagram. Any signal that has excessive routing should be pointed out to the design engineer for evaluation.

It is important for the designer and the design engineer to work closely on the chip plan. Proper partitioning of the logic will reduce the number of tunnels and the parasitic routing capacitance of signal lines.

10.4 TIME CONSTANTS

The time constant of an interconnect line is complicated and difficult to determine accurately. The field capacitance is distributed and gate capacitance tends to be localized. For our purposes we will lump together all the capacitance and all the resistance and multiply the two values to find the time constant for the line.

Critical signal nodes require small time constants. In microprocessor-type designs, the data and address bus lines are considered critical. When preparing the chip plan, these lines require special attention.

10.5 PRIORITY OF SIGNALS

The design engineer should establish the priority of signals. Critical signals will have the highest priority. When two signal lines are in conflict on a layer such as metal, the line having the lower priority will receive the tunnel. On other layers, the signal having the highest priority will be tunneled to the layer that will receive the least capacitance and resistance. (This may mean that there is no tunnel.)

When signals are in conflict with power or clock buses, the power or clock will remain on metal unless an exception is permitted by the design engineer, such as in the case of a supercritical signal line.

10.6 PRIORITY OF POWER BUSES

Normally, V_{SS} will have the higher priority of the two buses, V_{DD} and V_{SS}. Exceptions to this rule will be detailed by the design engineer. Separate buses are generally used for output buffers and internal logic.

**10.7
ION MIGRATION**

Ion migration occurs when a metal line develops voids or clumps due to electrons striking the metal molecules. Ion migration is affected by temperature and current density. The composition of the metal is also a factor, as this is an industry standard value for most processes. The current in a metal line will be kept to 1 mA/μm or less in this design. Rules covering current density vary from process to process.

**10.8
FUSING**

Fusing of lines seldom occurs internal to the die. Fusing does occur on and around input and output lines as a result of handling and abuse. Both metal and poly lines may fuse. Metal will fuse in contact regions. Fusing is not a concern as long as the designer abides by the design rules. Fusing has been used to program MOS read-only memories.

**10.9
ESTIMATING POWER
REQUIREMENTS**

The sizes of all transistors and power buses are the responsibility of the design engineer. The design engineer may require the designer to calculate the power on the bus lines using a transistor current chart provided by the design engineer. The engineer may give the designer current requirements for each cell and ask the designer to size the buses or the design engineer may size the buses. In either case, the power bus will be sized based on the estimated power requirements. Worst-case current is used to determine estimated power. Maximum current at the lowest operating temperature is normally used for worst-case current values.

One method for determining worst-case current drawn by the chip is to assume that all the loads are drawing maximum currents. The current computed is then divided by 2. A more accurate method requires knowledge of the logical states. The loads that draw current are used to compute the current consumed by the chip. The current requirement for output buffers is calculated separately from that for the rest of the chip.

11

CHIP PLANNING

The chip plan is a drawing that shows the placement of the logic as it is to be organized and located on a die or mask set. The element that has the greatest impact on die size is the chip plan. Frequently, the design engineer will partition the logic diagram by logical function. The engineer will arrange the logic diagram such that the logic is easy to follow. Often, the pads will fall in the same order as the package pins. The engineer may or may not consider the routing and optimization of the logic partitioning. Many designers follow the partitioning as it appears on the logic diagram without regard to the size or shape of the areas involved. Doing this can create an inefficient design.

The IC designer should organize the chip plan based on the routing of critical lines, the number of lines of interconnect, the logic size, the power bus, and the bonding pads. Some items, such as large blocks of ROM and RAM, will often dominate the chip plan to such an extent that they will dictate chip size. All other random logic will fill in around the memory arrays. Random logic can frequently be organized to reduce interconnections between elements within as well as between different blocks. Sometimes, rotating a cell can reduce bus routing to nothing. Ideally, there should be no signal buses on a die. One piece of logic should fit right next to the other.

**11.1
ORGANIZING
THE LOGIC**

The logic should be organized in such a manner as to reduce interconnections. The design engineer probably has organized the logic by function. The organization by reduced interconnections or logic function may not be the same. Sketches of cells may be helpful in determining the actual area required for a piece of logic. Serious study should be conducted. A multi-input NOR gate might be implemented as a single line instead of many.

**11.2
ARRANGING THE PADS**

Frequently, the locations of pads are proposed in a specification. Often, the pads may be relocated if there is sufficient justification. If, after serious study of the logic, it appears that there are better locations for the pads, the design engineer should be consulted.

**11.3
PARTITIONING LOGIC**

Partitioning of the logic on the diagram is probably done as a result of function. On the chip, such factors as area, aspect ratio, busing, routing of signals, and so on, must be considered. This is what the chip plan is for. At this stage, the actual implementation of the logical gates and device size must be considered. All routing should be kept to a minimum. Most designs have over 60% of their die area in interconnections. A good design will produce considerably less than this.

**11.4
ROUTING OF CLOCKS
AND POWER BUSES**

Clock and power buses should be routed such that all the logic will receive power and that gates that require a clock have it. These lines should be kept as short as possible. Power buses occupy considerably more area than that needed for any other line. Crossovers and tunnels should be minimized.

**11.5
ESTIMATING LOGIC SIZE**

Experience, history, and sketches are used to estimate the area that a piece of logic will occupy. Short of actually drawing the gates, transistors, and interconnections, there is no real way to determine the actual area to allocate. Best-guess estimates are obtained from similar designs, corrected for differences in device sizes and logic. When in doubt, allow for extra room.

Remember, power buses occupy the metal layer only. Logic may be built beneath the metal in these areas. Utilize them fully.

**11.6
CHIP SIZE ESTIMATES**

Prior to designing a part, the design engineer needs to know how big it will be. The initial estimates are made based on the complexity of the logic and history of past designs. As the design of the logic progresses, more details are learned. New chip sizes are currently estimated based on average sizes of certain common gates. This provides a reasonable estimate if interconnect and pads and scribe are considered. At this point a rough chip plan is formulated.

Once the designer receives the logic diagram with device sizes annotated, chip planning can begin.

12

SCRIBE LINE, BONDING PADS, PACKAGING, ALIGNMENT KEYS, END-POINT DETECTORS, TEST SITES, AND LOGOS

The processing and manufacturing requirements necessary to produce a proper die require features other than the logic. Scribe lines and bonding pads are needed in order to package the die. The layers require logos. Alignment keys and end-point detectors are required to process the die properly. Test sites are used to determine if the electrical properties are right.

The requirements for these important ancillary features are fixed and are detailed in the design rules. Deviations must be approved not only by the design engineer, but also by all other groups involved.

12.1 PAD SIZES AND ARRANGEMENT

The position of pads should be such that each pad is near its bonding post. The bonding wires that go from the pad to the post should be short. Under no circumstances should the bonding wires cross. The design engineer will specify the order of the pin-out and the package to be used.

Pad sizes are specified in the design rules. If the die is to be optically shrunk, the pads must be made larger so that the end-product bonding pad sizes are in agreement with the design rules. Radial lines from the center of the die to the pads illustrate where the bonding pads should be located for the best bonding configuration.

**12.2
ROTATIONAL
ALIGNMENT**

Special alignment keys are placed outside the scribe line. These keys are used in the mask shop for checking the size of the die and for measuring rotational alignment. These keys are never seen on the finished product or on the masks. They may or may not appear on the composite drawing. On the circuit are a set of alignment keys which are used to align a mask with previous masking operations. These keys are used in measuring and correcting rotational alignment at the masking operation stage. These alignment keys are also used when checking mask contact.

An interesting note: Rotational alignment does not change with magnification. X and Y alignments do change. The rotational alignment keys control angular positioning of the die during mask fabrication.

**12.3
STEP-AND-REPEAT**

Step-and-repeat is a process in which the reticle is exposed repeatedly on a plate to create a mask. The reticle is the exposed plate, usually at 10X magnification, which represents the shapes that will appear on the mask for a die.

A process monitor or test chip may be "stepped" onto the mask in several spots. Process monitors frequently appear at the top center and bottom left and right on a wafer.

The product produced by step-and-repeat is called a *master*. When contact printing is used to process the die, submasters and working plates are made from the master.

**12.4
PROCESS CONTROL
TECHNIQUES**

To ensure that the wafers are being processed properly, measurements are made at predetermined points in the process flow. Critical dimensions are measured and used to verify that line width and spaces are correct. Certain geometries on the test site are used. To check etches, the scribe line or special shapes in the scribe lines are used. Coarse alignments for masking operations use the scribe line. Fine alignments use the alignment keys.

**12.5
LOGOS**

A manufacturer's trademark or logo is placed on each die. Lawyers have strong feelings about the appearance and the placement of logos on designs. The logo helps protect the company designing a circuit from design theft.

**12.6
MASK NUMBERS**

To identify each mask set, a mask number is placed on each chip designed. Changes in single layers are given revision letters. Some manufacturers require revision letters on the composite drawing. Other manufacturers place the revision letters on the mask label. Changes in several levels cause some manufacturers to change the mask set to avoid confusion.

**12.7
ENGINEERING
CHANGE ORDERS**

Engineering change orders (ECOs) are part of the job. Any change to instructions or documents should be completely documented. One method of doing this is to use an engineering change order. All documents affected should be appropriately updated. The revision block should be annotated and the ECO referenced.

13

INPUT PROTECTION, CHANNEL STOPS, AND GUARD RINGS

Input protection, channel stops, and guard rings are used to protect the internal logic from outside noise and stress.

13.1 STATIC PROTECTION SCHEMES

Shipping, handling, testing, and installation of the MOSFET IC subjects the component to electric static discharge (ESD). The static input protection device is designed to prevent serious damage to the device as a result of this exposure. It is possible to cause the gate oxide on a semiconductor to rupture by placing the oxide in a strong electric field. Most gate oxide ruptures occur when an input pin is subjected to a static electric voltage that discharges through the pin. The input protection device as specified in the design rules assumes that V_{ss} is grounded. The details are described in the design rules as to the proper implementation of this device. Any deviation may cause total failure of the protection scheme.

Protection devices are evaluated by charging a 100- to 150-pF capacitor to a voltage and discharging it through a 1000- to 1500-Ω resistor into the pins of the device (Figure 13.1). This is repeated several times for each voltage. The voltage is increased and the test is repeated until the part fails. A good protection device will withstand 2000 V.

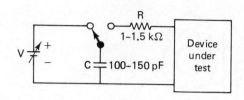

FIGURE 13.1 Schematic for the test fixture to evaluate input protection networks.

13.2 FIELD INVERSION

Field inversion occurs when a metal or poly line passing over a field region causes the substrate to invert. If the field inverts, parasitic transistors are created and leakage paths occur. These leakage paths may cause the part to malfunction. In this process, voltages equal to or less than 7 V will not cause field inversion. These voltages are within the normal operating levels. Bootstrap buffers, bias generators, and lines such as clocks from off-chip devices have voltages in excess of 7 V.

Although field inversion may occur, the damage due to leakage paths may be inhibited. This may be done by using a poly line beneath the offending metal line. The poly line may have a bias to neutralize the field on the substrate. This will block the leakage path. N^+ may be used either as a getter or to nullify a leakage path. The N^+ would be placed beneath the metal line. An N^+ region used around a gate for this purpose is known as a *guard ring*. Guard rings are used extensively in CMOSs and in designs where the applications require voltages higher than normal design levels for the process.

13.3 LATCH-UP

Although latch-up does not occur in NMOS circuits, it is introduced here because of its importance in CMOS layout. *Latch-up* is a term used to describe SCR (silicon-controlled rectifier) action in the PNPN or NPNP diode which occurs in the CMOS process. This diode is located along the surface of the die and is associated with the well (Figure 13.2a). Schematically, latch-up appears as shown in Figure

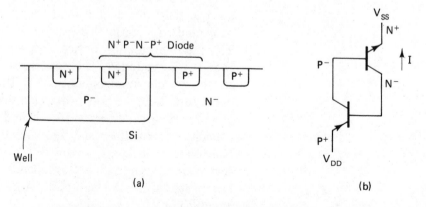

(a)

(b)

FIGURE 13.2 CMOS silicon controlled rectifier (SCR) structure.

13.2b. Latch-up can be reduced and eliminated by proper strapping of the well, preventing voltage drops from occurring in the well.

13.4 CHARGE INJECTION

Forward-biasing the N^+ junctions in the P^- substrate allows current to enter the substrate. The additional carriers can combine with stored charges in the logic elements, causing failure of the part. Usually, the injection tends to localize near the point of entry.

Undershooting signal lines such as clocks tends to cause the most problems in applications. Improperly designed tests can take their toll on parts by injecting a charge into the substrate by the forward biasing of junctions.

14

EXAMPLE: THE CHIP PLAN

**14.1
PURPOSE**
The design of a cell is dependent on how it will interface with the other cells, signal, and power busing that make up an integrated circuit. It is important for an integrated-circuit designer to understand what constitutes a chip plan, and how to use the plan to best advantage. This project is included to assist the student in developing a chip plan for the design project and to become familiar with the ingredients of a good plan.

**14.2
INGREDIENTS OF A
GOOD CHIP PLAN**
A good chip plan will contain:

1. The approximate location of the bonding pads necessary to connect the circuit to the package leads
2. The planned routing of the major power buses
3. The anticipated location of logical functions
4. The routing for critical signals and those which are in common with a multitude of functions

14.3

DEVELOPMENT
OF THE PLAN

In planning the organization for an integrated circuit, three items will be used:

1. The bonding requirements found in the device specification, usually in the form of a proposed bonding diagram

2. The logic diagram

3. Sketches made of the basic cells

Figure 8.5 contains a copy of the package bonding diagram. The pin-out requirements for the project are defined on the logic diagram as it appears in Figure 14.1. Sketches of the basic cells were part of the project of Chapter 9. Some of these cells and others used to make size estimates appear in Figures 14.2 to 14.5.

In developing the first cut at the chip plan, the proposed bonding and the package diagram are considered. Specific cuts are usually set forth regarding the placement of the bonding plans. The essence of the rules is that the pads should be spaced evenly around the edge and close to the package bonding posts. The distance from the bonding pad on the integrated circuit to the bonding posts on the package should be as short as possible (see Figure 14.2).

The layout design rules in Chapter 7 define the bonding pad size and spacing. The closest that a pad may be located to another pad is 138 μm. The minimum pad size is 138 μm. The first pass at developing the chip plan will be the placement of the pads (see the chip plan in Figure 14.6). This is done to verify that the pin count agrees with the package and that V_{SS}, V_{DD}, and critical signals such as the clock signals are on appropriate pins.

The next step is to estimate the area required to implement the logic. This is done using empirical methods. That is, sketches are made of certain cells and estimates are made of certain cells, and estimates are derived based on these sketches (see Figures 14.2 to 14.5). For common logical elements that are repeated, such as buffers, one sketch may be used to estimate the area for all. In the case of arrayed elements, unique elements such as bit cells, input buffers, and loads may be sketched. The estimated area to be used by the array may be based on that (see Figures 14.1, 14.4, and 14.6).

In the case of random logic, a single gate may be drawn which represents the average type of gate. Device sizes and the number of inputs and outputs are considered. In our example, the gate area is calculated separately from the coupler area (see Figure 14.5). The logic diagram is sectionalized. The gates in each section are totaled. The area is then figured for each section (see Figure 14.7).

Figure 14.6 represents the chip plan for the project, and the pin and pad assignments are illustrated. The buffer cell estimates are shown with dashed lines. Other logic is shown as boxes. The V_{DD} bus is shown as crosshatched lines tilted to the left, and the V_{SS} power bus is shown tilted to the right. Note that the critical path is routed

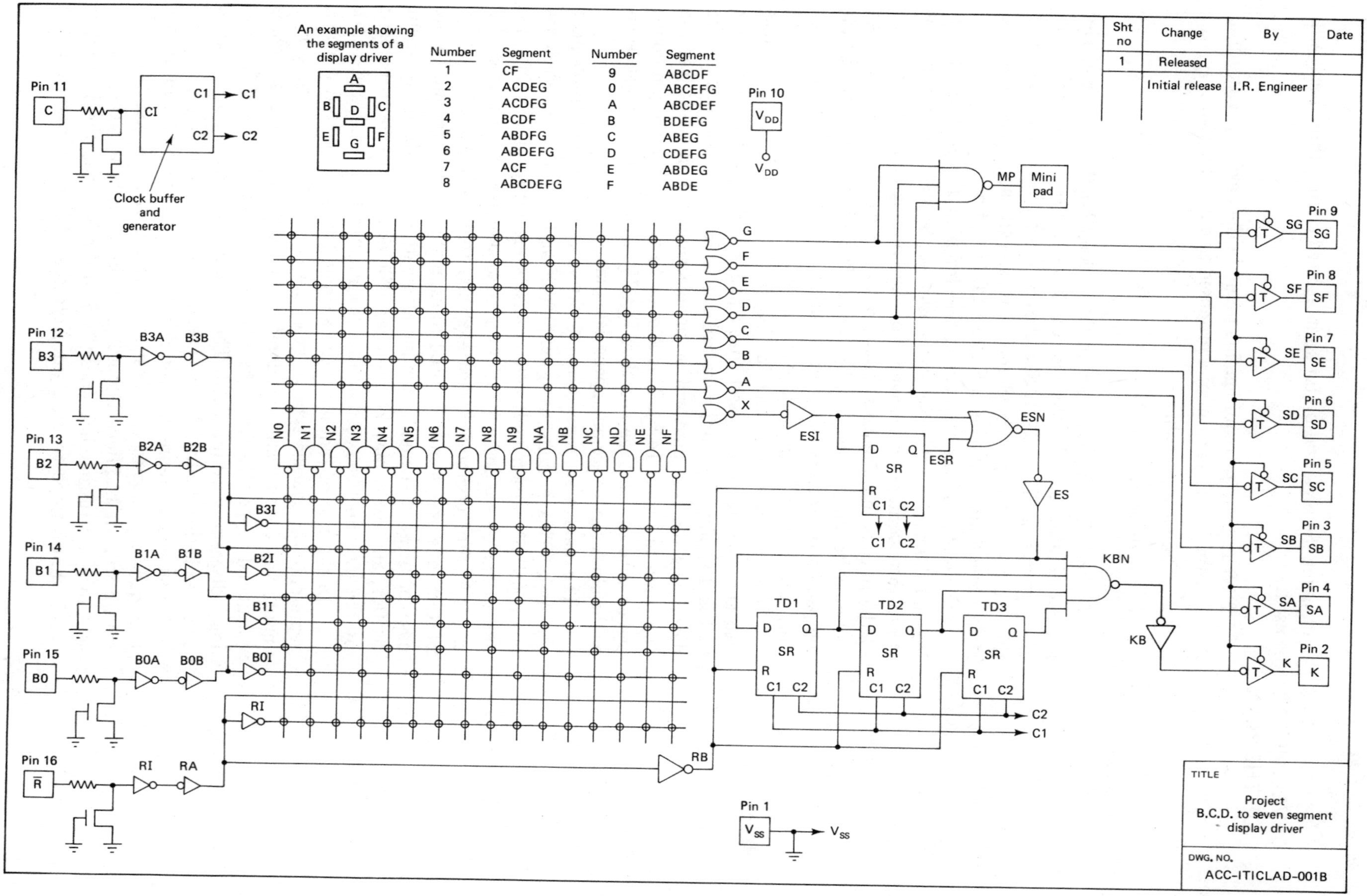

FIGURE 14.1 Logic diagram.

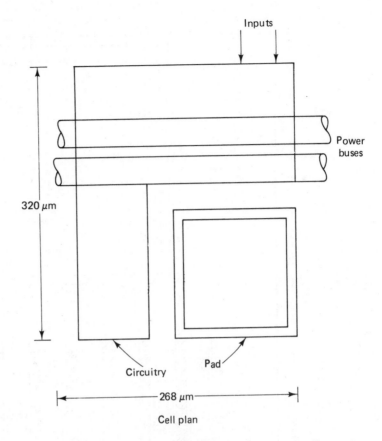

FIGURE 14.2 Estimating area for S output buffers.

so as not to cross the power bus and is shown as having no tunnels or crossovers. Power is bused to the output buffers on a bus separated from the internal logic. The power bus is routed to each cell without tunnels. Note that the drawing is drawn to approximate scale dimensions that permit a chip size to be estimated. The size of this chip plan is 2080 by 2080 μm. Shown also are the logo, alignment keys, mask number, and test sites. Fiducial marks are not shown.

Now that the integrated-circuit logic implementation has been organized and the plan implemented, it is possible to lay out the final version of the cells and to develop the complex cells. The design engineer can now specify bus sizes and prepare more accurate device sizes. Bus sizes will be determined by the current requirements and the logic attached to the bus. The length of critical nodes and the load requirements may be estimated. The chip plan will be critiqued by the design engineer. If there are problems or suggested improvements to the chip plan, these will be brought up at this time. Upon completion and approval of the plan, the integrated-circuit designer is ready to start the design.

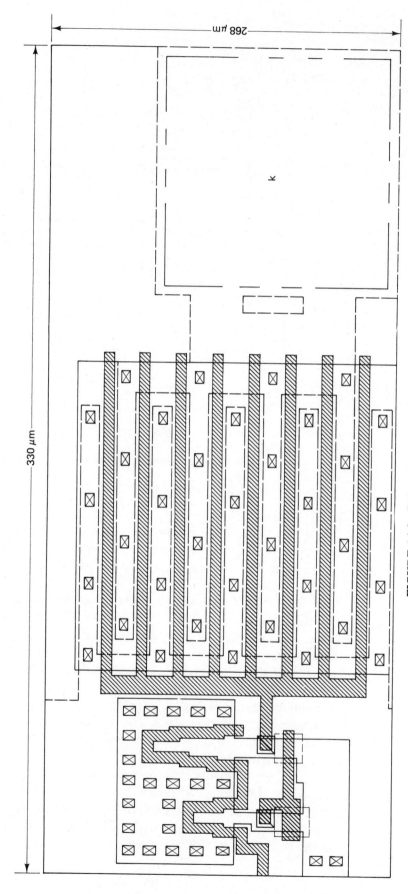

FIGURE 14.3 Estimating area for K output buffer.

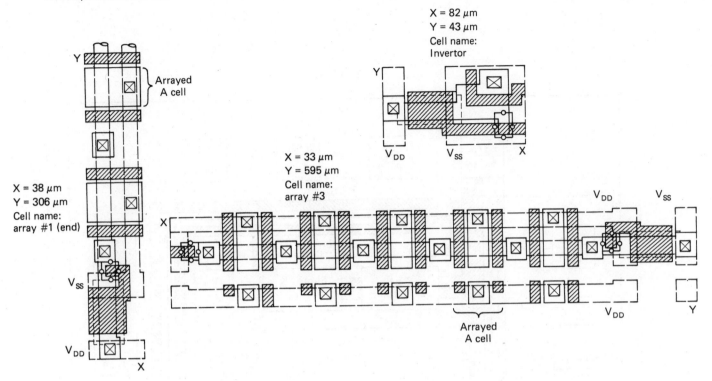

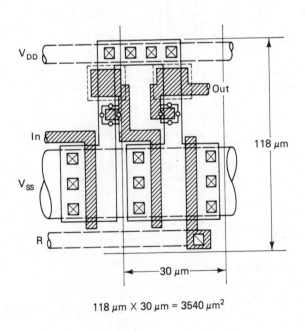

FIGURE 14.4 Estimating area for logic arrays.

FIGURE 14.5 Estimating area for random logic.

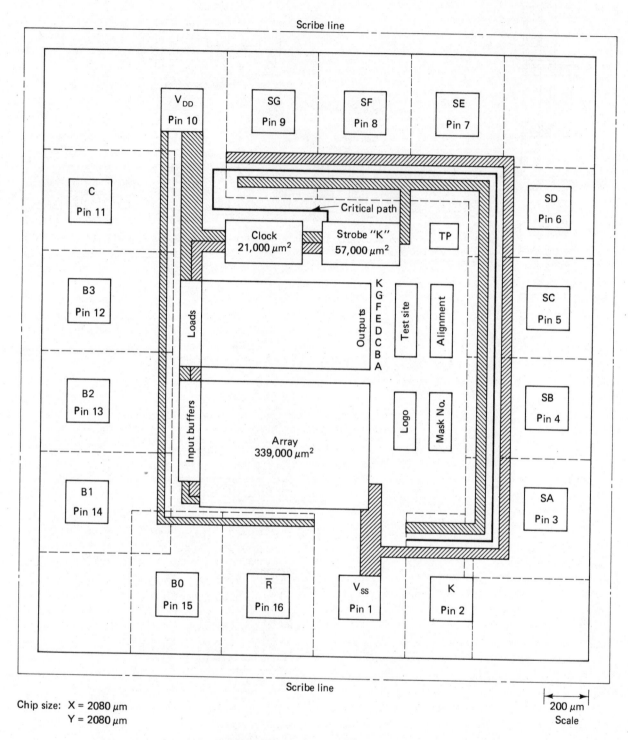

Scribe line

V_{DD}
Pin 10

SG
Pin 9

SF
Pin 8

SE
Pin 7

Critical path

Clock
21,000 μm²

Strobe "K"
57,000 μm²

TP

C
Pin 11

SD
Pin 6

Loads

Outputs

Test site

Alignment

B3
Pin 12

K G F E D C B A

SC
Pin 5

Input buffers

Logo

Mask No.

B2
Pin 13

SB
Pin 4

Array
339,000 μm²

B1
Pin 14

SA
Pin 3

B0
Pin 15

R̄
Pin 16

V_{SS}
Pin 1

K
Pin 2

Scribe line

Chip size: X = 2080 μm
Y = 2080 μm

200 μm
Scale

FIGURE 14.6 Chip plan.

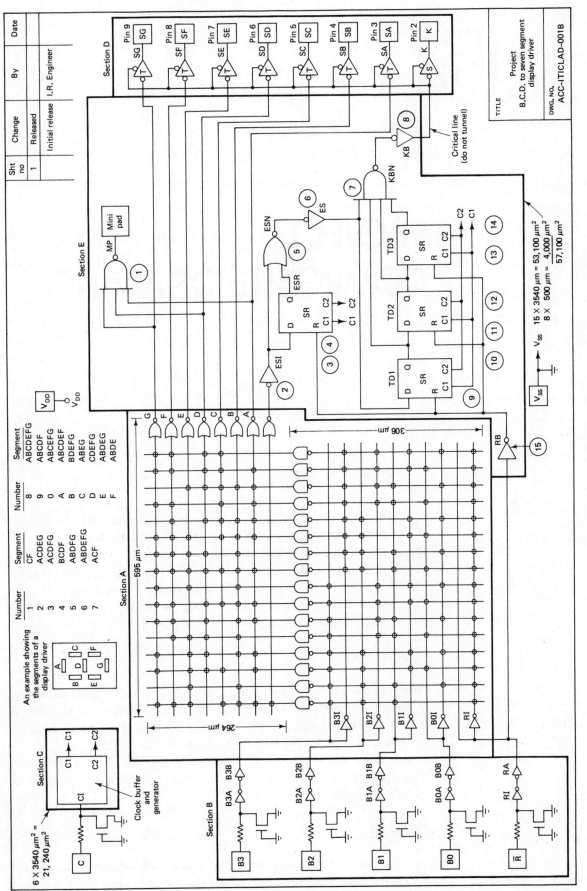

FIGURE 14.7 The logic diagram blocked into logic section to assist in die size and chip plan estimates.

15

PROJECT: COMPOSITE DRAWING

Each cell drawing is actually a composite drawing. All the various mask layers that come into play to make up the elements of the cell are shown. The large cells that make up the different sections of the overall chip are called *section sheets*. Different smaller cells will be placed on the section sheets. All the section sheets and the cells associated with the section sheets comprise what is called the *composite drawing* (see Figure 15.1).

Purpose

The purpose of this project is to develop the techniques necessary to design complex and special-purpose cells. The student will design two types of output buffers, a static shift register, complex random logic, a clock buffer, and a PLA. Appendix A contains the cell description for these gates. After designing the cells, the student will then place these cells with the ancillary composite elements to form a small LSI chip.

Procedure

1. The student will draw and place each cell onto the appropriate section sheet in such a manner that the logic that

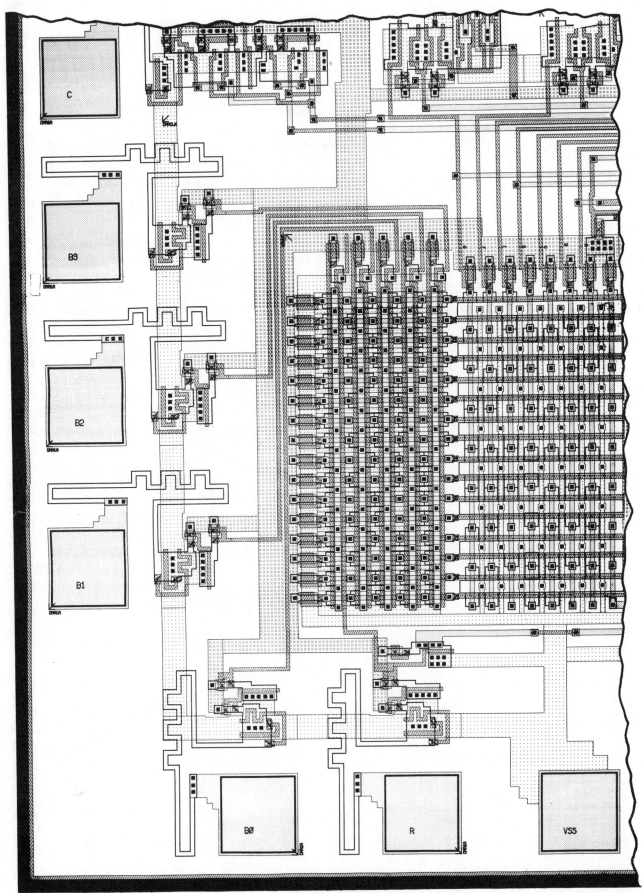

FIGURE 15.1

appears on the logic diagram can be implemented. The chip plan will be used to assist in the drawing and placing the cells.

2. The student will then draw in the necessary interconnects and line routing to implement the logic and connect the power.

3. The student will then place the alignment keys, test sites, and etch keys. A logo and the mask set name will be placed. Students may use their initials for the mask set name.

4. The student will draw in the details for the scribe-line special notes and fiducial marks.

5. After the composite has been finished, the student will work with another person and verify the correctness of the logic, device sizes, and design rules.

The procedure in developing the cells is not too different from that of the simple cells. The logic must be broken down into schematics. Since these cells interface with other circuitry, a review of the chip plan is mandatory. In the case of the output buffers, the design of the cell is dependent on the location of the pads with respect to the power buses, signal lines, other circuitry, and other pads and buffers. Output buffers will "sink" or "source" current to the outside world through the pads. Where the current requirements are high, techniques such as metal strapping will be required to reduce voltage drops in the N^+ source-drain nodes of the transistor. Metal strapping is often used in high-powered clock buffers to reduce propagation delay by reducing resistance of the N^+ nodes. When making sketches of the cells, allow for straps on the source and drain nodes of the output transistors for the clock generator and the output buffer.

A close study of the clock generator, shift register, and random logic will reveal a complex interconnection which is easily solved by the use of jumpers. When planning these cells, allow sufficient room between the power buses for routing the clock lines on metal.

APPENDIX

A

DESIGN PROJECT CELL BOOK

BCD-to-Seven-Segment Display Driver

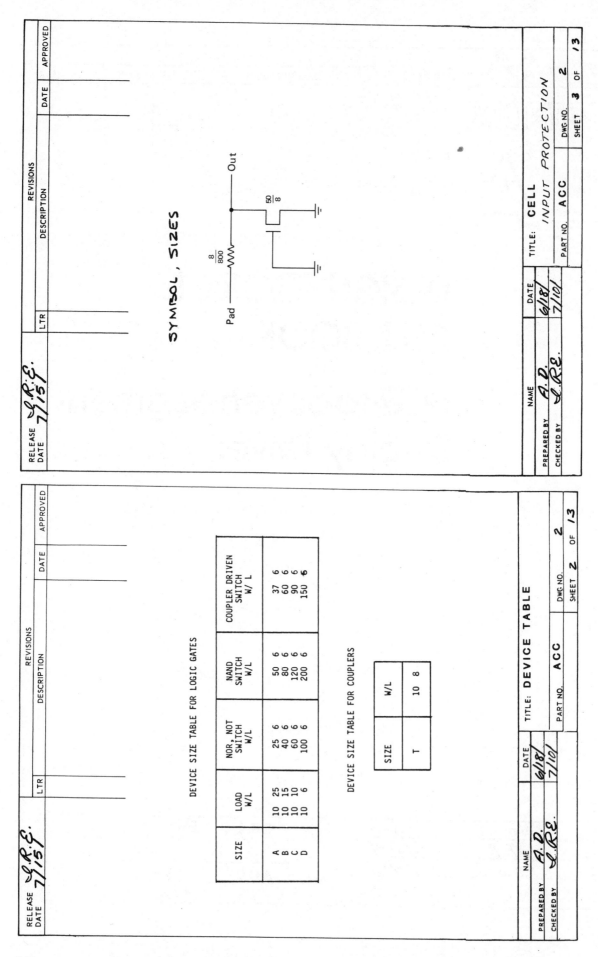

Top frame

SYMBOL, SIZES

Pad — $\dfrac{8}{800}$ —●— Out

$\dfrac{50}{8}$

	NAME	DATE
PREPARED BY	A.D.	6/18/
CHECKED BY	L.R.E.	7/10/

TITLE: **CELL** INPUT PROTECTION

PART NO. **ACC**	DWG. NO. **2**
	SHEET **3** OF **13**

Bottom frame

DEVICE SIZE TABLE FOR LOGIC GATES

SIZE	LOAD W/L	NOR, NOT SWITCH W/L	NAND SWITCH W/L	COUPLER DRIVEN SWITCH W/L
A	10 25	25 6	50 6	37 6
B	10 15	40 6	80 6	60 6
C	10 10	60 6	120 6	90 6
D	10 6	100 6	200 6	150 6

DEVICE SIZE TABLE FOR COUPLERS

SIZE	W/L
T	10 8

	NAME	DATE
PREPARED BY	A.D.	6/18/
CHECKED BY	L.R.E.	7/10/

TITLE: **DEVICE TABLE**

PART NO. **ACC**	DWG. NO. **2**
	SHEET **2** OF **13**

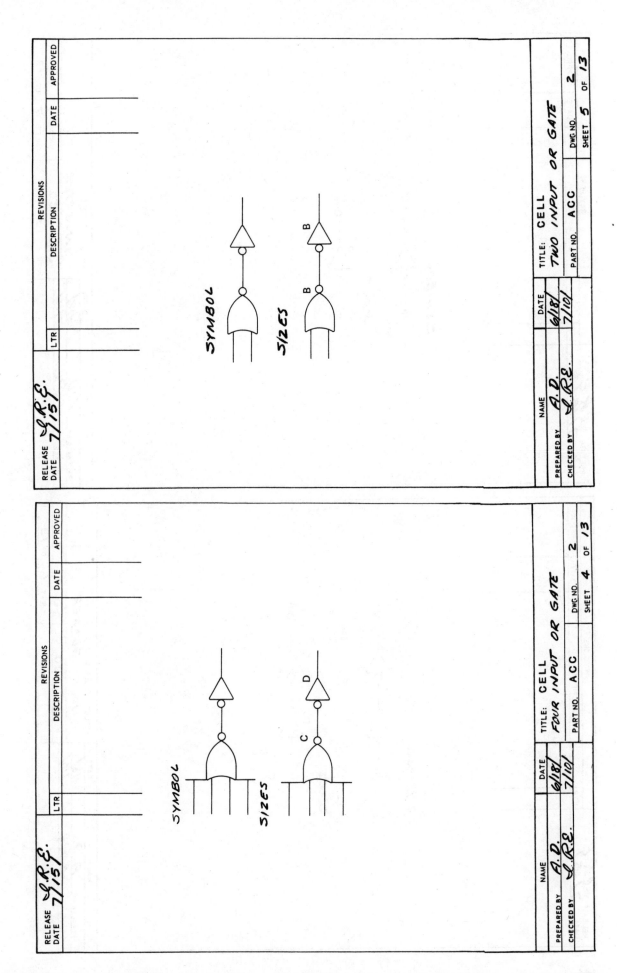

TITLE: CELL
TWO INPUT OR GATE
PART NO. ACC
DWG. NO. 2
SHEET 5 OF 13

TITLE: CELL
FOUR INPUT OR GATE
PART NO. ACC
DWG. NO. 2
SHEET 4 OF 13

SYMBOL

SIZES

SYMBOL

SIZES

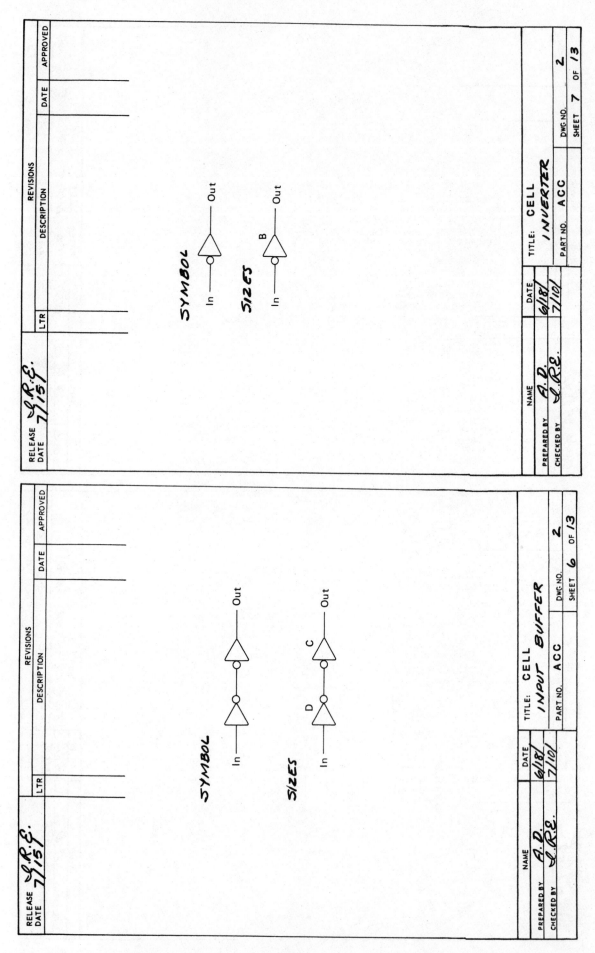

SYMBOL

In ▷ Out

SIZES

B
In ▷ Out

RELEASE DATE J.R.E. 7/15/

REVISIONS
LTR | DESCRIPTION | DATE | APPROVED

NAME | DATE
PREPARED BY A.D. | 6/18/
CHECKED BY J.R.E. | 7/10/

TITLE: CELL INVERTER
PART NO. ACC | DWG. NO. 2
SHEET 7 OF 13

SYMBOL

In ▷○─▷ Out

SIZES

D
In ▷○─▷ C Out

RELEASE DATE J.R.E. 7/15/

REVISIONS
LTR | DESCRIPTION | DATE | APPROVED

NAME | DATE
PREPARED BY A.D. | 6/18/
CHECKED BY J.R.E. | 7/10/

TITLE: CELL INPUT BUFFER
PART NO. ACC | DWG. NO. 2
SHEET 6 OF 13

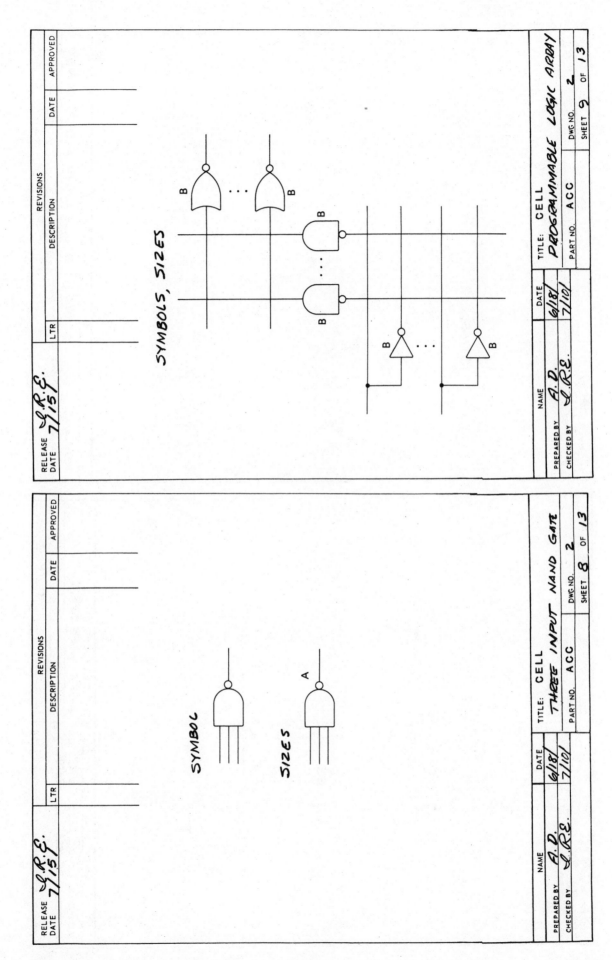

SYMBOLS, SIZES

SYMBOL

SIZES

RELEASE
DATE 7/15/

REVISIONS

LTR | DESCRIPTION | DATE | APPROVED

TITLE: CELL
PROGRAMMABLE LOGIC ARRAY

NAME | DATE
PREPARED BY A.D. 6/18/
CHECKED BY L.R.E. 7/10/

PART NO. ACC
DWG.NO. 2
SHEET 9 OF 13

RELEASE
DATE 7/15/

REVISIONS

LTR | DESCRIPTION | DATE | APPROVED

TITLE: CELL
THREE INPUT NAND GATE

NAME | DATE
PREPARED BY A.D. 6/18/
CHECKED BY L.R.E. 7/10/

PART NO. ACC
DWG.NO. 2
SHEET 8 OF 13

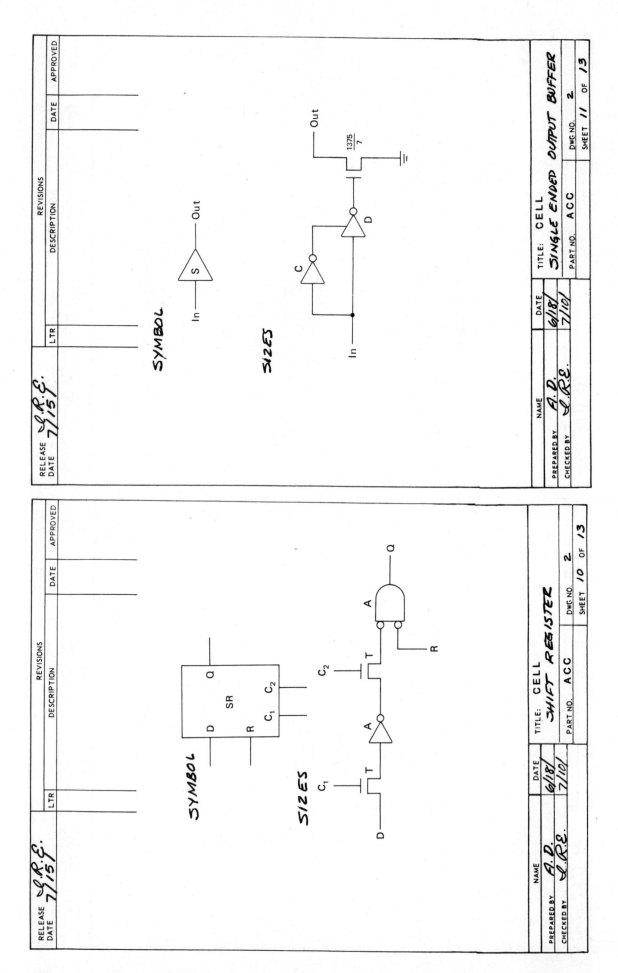

SYMBOL

In —▷ S — Out

SIZES

C

D

In

Out

$\frac{1375}{7}$

TITLE: CELL
SINGLE ENDED OUTPUT BUFFER

	NAME	DATE
PREPARED BY	A.D.	6/18/
CHECKED BY	L.R.E.	7/10/

PART NO. | ACC | DWG. NO. | 2

SHEET 11 OF 13

SYMBOL

D | SR | Q
R | | C₁ C₂

SIZES

C₂

C₁

D

A

T

T

A

R

Q

TITLE: CELL
SHIFT REGISTER

	NAME	DATE
PREPARED BY	A.D.	6/18/
CHECKED BY	L.R.E.	7/10/

PART NO. | ACC | DWG. NO. | 2

SHEET 10 OF 13

146

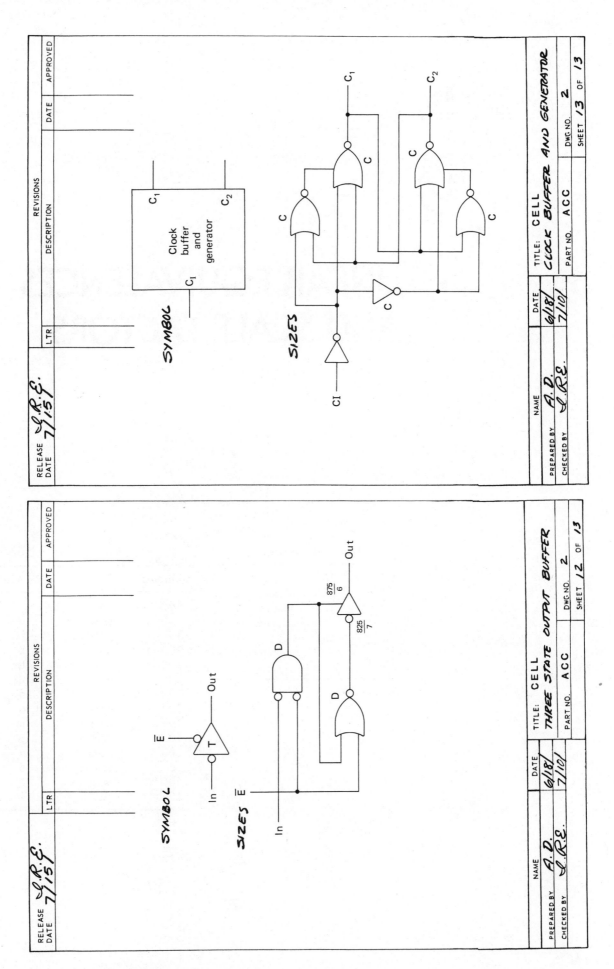

Top drawing (Sheet 13):

REVISIONS

LTR	DESCRIPTION	DATE	APPROVED

RELEASE DATE $\mathcal{L.R.E.}$ 7/15/

SYMBOL

Clock buffer and generator

C_1

C_2

C_1

SIZES

C

C

C

C

C

C

C_1

C_2

CI

	NAME	DATE
PREPARED BY	A.D.	6/18/
CHECKED BY	L.R.E.	7/10/

TITLE: CELL
CLOCK BUFFER AND GENERATOR

PART NO. ACC   DWG. NO. 2

Bottom drawing (Sheet 12):

REVISIONS

LTR	DESCRIPTION	DATE	APPROVED

RELEASE DATE $\mathcal{L.R.E.}$ 7/15/

SYMBOL

$\overline{E}$

In   T   Out

SIZES

$\overline{E}$

In

D

D

$\frac{875}{6}$

$\frac{825}{7}$

Out

	NAME	DATE
PREPARED BY	A.D.	6/18/
CHECKED BY	L.R.E.	7/10/

TITLE: CELL
THREE STATE OUTPUT BUFFER

PART NO. ACC   DWG. NO. 2

APPENDIX

LINEAR EQUIVALENCES AND SCALE FACTORS

Linear Equivalences[a]	
1 mil	10^{-3} inch
	25.4 micrometers (μm)
	10^{-10} meter (m)
	10^{-8} centimeter (cm)
1 Å	10^{-4} μm
	10^{-1} nanometer (nm)
1000 Å	0.1 μm

[a]Å, angstrom.

Scale Factors		
Multiplier	Prefix	Abbreviation
10^{12}	tera	T
10^{9}	giga	G
10^{6}	mega	M
10^{3}	kilo	k
10^{2}	hecto	h
10	deka	da
10^{-1}	deci	d
10^{-2}	centi	c
10^{-3}	milli	m
10^{-6}	micro	μ
10^{-9}	nano	n
10^{-12}	pico	p
10^{-15}	femto	f
10^{-18}	atto	a

APPENDIX C

GLOSSARY

A/D converter Analog-to-digital converter. A device to convert variable or analog signals to digital representation.

Access time The interval between a request for stored information and the delivery of the information; often used as a reference to the speed of a memory. Read time is access time required to extract information. Write time is time required to store information. Since the access time to different locations (addresses) of memory may be different, access time in a memory is the path that takes the longest time.

Accumulator A data-word holding register for arithmetic, logical, and input/output operations. Generally, an accumulator is a register used in arithmetic circuits, such as the totalizing circuit of an adder, or as a place to hold one operand for arithmetic and logical operations.

Accuracy The deviation, or error, by which an actual output varies from an expected ideal or absolute output. Each element in a measurement system contributes errors, which should be specified separately if they contribute significantly to the degradation of total system accuracy. In an analog-to-digital converter, accuracy is tied to resolution; an 8-bit A/D, for example, can resolve to 1 part in 256 ($2^8 = 256$), so the best accuracy as a percentage of full-scale range is theoretically 1/256, or about 0.4%.

Acoustic coupler A device that receives digital data, converts them to tones suitable for transmission over telephone lines, then couples the tones into a telephone line; it performs the reverse operation as well. Usually in the form of a cradle, into which a telephone handset is placed.

Active elements Those components in a circuit which have gain or in which direct current flows, such as diodes and transistors.

Active state The state in which the logic circuit performs its particular function. For example, the active state of the power set (initializer) circuit provides a reset function when power is initially applied, therefore, the logic levels should be shown for this active state.

Activity The presence of active signals. These signals are then detected in the activity checker, which generates an activity signal.

Adder Switching circuit that generates sum and carry bits.

Address A code designating the particular unit (vehicle, decoder, teleprinter, radiotelephone, etc.) that must respond to an incoming message. When used with computers, it is a code that designates the location of information or instructions in the main storage, peripherals, etc.

Alignment The arranging of the mask and wafer in correct positions, one with respect to the other. Special alignment patterns are normally part of the mask.

Alignment key A group of shapes used to ensure the proper positioning of different layers on a chip.

Ampere (A) The steady electrical current produced by 1 volt applied across a resistance of 1 ohm.

Analog computer A continuous-variable computer, or nondigital computer. A differential analyzer. Measures the effect of changes in one variable on all other variables in a system. Its operation is analogous to that of a slide rule.

Analog switch A device that allows transfer of signals in both directions whenever the switch is enabled. When the switch is not enabled, it offers a high impedance.

And gate *See* Gate, AND.

Angstrom One hundred-millionth of a centimeter. A unit used to measure the wavelength of light. Oxide films used in the manufacture of metal-oxide semiconductors are measured in angstroms. The symbol for the angstrom is Å. (1 mil = 254,000 Å.)

Arithmetic logic unit The section of a central processing unit (CPU) that makes arithmetic and logical comparisons and performs arithmetic functions (adds, subtracts, shifts, ANDs, ORs, etc.)

ASCII A widely used code (American Standard Code for Information Interchange) in which alphanumerics, punctuation marks, and certain special machine characters are represented by unique, 7-bit binary numbers; 128 different binary combinations are possible ($2^7 = 128$), thus 128 characters may be represented. (A 6-bit ASCII code is also available in which 64 characters may be represented.) Also a protocol.

Aspect ratio The aspect ratio of a metal-oxide-semiconductor transistor is its width divided by its length. The aspect ratio of an integrated circuit is its smallest dimension divided by its longest dimension.

Assemble A procedure used by programmers to convert programs drawn up in symbolic form (generally flowcharts) into a computer language.

Assembler A software program that translates assembly language into machine language.

Assembly The translation of mnemonic code into machine language by an assembler.

Assembly language A mnemonic programming language that approximates machine language. It saves the programmer the trouble of remembering the bit patterns in instructions.

Assembly listing A printed list made by the assembler to document an assembly language program.

Assert To cause a control signal to *assume* its functional logic level. For example, when the RESET signal is asserted (in positive logic), it assumes a logical low; however, when the AS (address strobe) signal is asserted (in positive logic), it assumes a logical high.

Astable multivibrator A free-running electronic circuit that generates pulses which can be used as timing signals or other similar signals.

Asynchronous Operation of a switching network by a free-running signal that triggers successive instructions; the completion of one instruction triggers the next.

Asynchronous communications interface adapter (ACIA) *See* Universal asynchronous receiver-transmitter (UART).

Automated bonding A method of attaching bonding wire to a chip by means of a machine.

Back bias Voltage applied to the substrate of an integrated circuit to enhance certain characteristics or performance.

Bar An integrated circuit. "Bar" is a term used to describe a part of a wafer. Junction field-effect transistors were made from tiny bars of silicon. Other terms, such as "die" or "chip," may be used interchangeably with "bar."

Beta ratio The current-carrying capabilities of a switch versus the current-carrying capabilities of the load.

Binary A system of numerical representation that uses only two symbols, 0 and 1. In computers, "binary" refers to a number of systems involving only two possibilities (typically high or low, positive or negative, presence or absence of pulses). In mathematics, "binary" refers to a number system with a base of 2. Each digit position represents a power of 2, rather than 10 as in the decimal system.

Binary-coded decimal (BCD) Four bits of binary information are used to encode one decimal digit. When a decimal digit is encoded in this way, it is called a binary-coded decimal (BCD).

Binary counter A circuit using a group of flip-flops to convert a series of pulses into binary form. *See also* Counter, binary.

Binary point The fractional dividing point of a binary numeral; equivalent to decimal point in the decimal numbering system.

Binary program A program (or its recorded form) in which all information is in binary microcomputer language. *See also* Machine language.

Bistable *See* Flip-flop.

Bit A unit of computer information equivalent to the result of a choice between two alternatives (1 or 0). Abbreviation for "*Bi*nary dig*it*."

"Blackbox" A description used for an electronic circuit which concerns itself only with the input and output, and ignores the interior elements, discrete or integrated.

Blow back A 100× photocopy of mask layers used to detect defects in a chip.

Bonding pad A large geometric shape made on the metal bar which is used to attach a wire from the chip to the package leads.

Bootstrap Using its own action to initiate or sustain itself: for example, a bootstrap operation (program) that the computer uses to load itself. Usually used to reconfigure certain computer areas during initialization, after power loss, etc.

Branch instruction An instruction that directs the microcomputer to leave the basic program at some point and branch to another point in the same program.

Breadboard An experimental arrangement of an electronic circuit to test feasibility. At one time prototype circuits were built on small boards resembling breadboards.

Buffer A noninverting stage, such as an emitter follower, which provides isolation and is sometimes used to handle a large fanout or to convert input and output levels. A temporary storage site (such as buffer storage, buffer register, etc.) that compensates for differences in data flow rates.

Bus transceiver A three-state device that provides two-way asynchronous data transfer. Data transfer capability is in either direction but not both simultaneously. The direction of data transfer depends on the logic level of the direction input signal.

Byte A group of 8 bits considered as an entity. Instruction sets for 8-bit microprocessors are defined in multiples of bytes. *See also* Nibble.

Capacitance The property of an electric nonconductor that permits the storage energy as a result of electric displacement when opposite surfaces of the nonconductor are maintained at a different potential.

Capacitor A device giving capacitance consisting of two conducting plates (separated by an insulator) oppositely charged.

Channel A region of surface conduction opposite in type from that of the bulk doping of an MOSFET transistor.

Channel length Distance between the drain and source regions of a MOSFET transistor.

Chip A single substrate on which all the active and passive elements of an electronic circuit have been fabricated utilizing the semiconductor technology. A wafer is processed with a multitude of integrated circuits. When the circuits are cut apart and are separated, they are called chips, bars, or dice.

Clear To restore a memory or storage device, counter, and so on, to a standard state, usually the "zero" state. Also called reset.

Clock A pulse generator that controls the timing of switching circuits in synchronous logic. Most digital VLSI components use at least one clock.

Clocked RS flip-flop The clocked RS flip-flop has two conditioning inputs which control the state to which the flip-flop will go at the arrival of the clock pulse. If the S (set) input is enabled, the flip-flop goes to the "1" condition when clocked. If the R (reset) input is enabled, the flip-flop goes to the "0" condition when clocked. The clock pulse is required to change the state of the flip-flop.

CMOS Complementary metal-oxide semiconductor. A logic family made by combining N-channel and P-channel MOS transistors.

COBOL A high-level programming language (Common Business-Oriented Language) for business applications.

Code To write a program; or, the program itself; often, used interchangeably with "language." A representation of characters as in ASCII coding. A system of symbols that can be used by a microcomputer/microprocessor and that in specific arrangements has a special external meaning.

Comparator A device to determine if two bits of information are in the same state (both 0 or both 1).

Complement The complement of a variable or function is the binary opposite of that variable or function. If a variable or function is 1, its complement will be 0. If a variable or function is 0, its complement will be 1. The complement of 011010 is 100101.

Concatenate To link together in a series or chain. For example, two 8-bit bytes can be concatenated to form one 16-bit word.

Conductor A substance capable of transmitting electricity.

Counter (a) A device capable of changing states in a specified sequence upon receiving appropriate input signals. (b) A circuit that provides an output pulse or other indication after receiving a specified number of input pulses. (Specific counters follow.)

Counter, binary A series of flip-flops having a single input. Each time a pulse appears at the input, one or more of the outputs or stages changes state; each successive stage represents increasing powers of the radix 2. The final counter output can either be a single output from the final stage or parallel outputs from each stage.

Counter, nonsynchronous (ripple) A number of series-connected flip-flops which divide (count-down) the input into a series output occurring at a lower rate. The actual division depends on the number of flip-flops and their specific connections. Each stage serves as a "clock" or input to the next higher stage.

Counter, programmable An integrated circuit containing series-connected flip-flops into which a binary number (within the counter range) can be programmed. The counter then counts the input pulses starting at the programmed number until the counter is full and resets. The cycle repeats.

Counter, ring A loop or circuit of interconnected flip-flops so arranged that only one is "on" at any given time and that as input signals are received, the position of the "on" state moves in sequence from one flip-flop to another around the loop.

Counter, shift A number of clocked series-connected flip-flops in which the flip-flops do not change with each clock. Instead, each flip-flop changes only once for each cycle of the counter. That is, in a three stage shift counter, the flip-flops change only once every three clocks, once every four clocks in a four-stage shift counter, etc.

Counter, synchronous A number of series-connected flip-flops in which the next state of each depends on the present state of the previous stage, and all state changes occur simultaneously with a clock pulse. The output (or equivalent count of input pulses) can be taken from the counter in parallel form.

CPU Central processor unit. That part of a computer that fetches, decodes, and executes program instructions and maintains status of results.

Crash A malfunction in hardware or software that requires the computer to be reset or restarted.

Critical dimensions A geometric shape of known size that is used to measure the size of specific geometries on various layers to determine process tolerances.

Cross-assembler A symbolic translator that runs one type of microcomputer to produce machine code for another type of microcomputer. *See also* Assembler.

Cross-under N regions or poly layers used to conduct signals under a metal line. The region is insulated from the metal by an oxide layer.

Cycle stealing A memory cycle stolen from the normal microprocessor operation for a DMA. *See also* Direct memory access.

Cycle time Timer interval at which any set of operations is repeated regularly in the same sequence.

Cycling tape When used by programmers, "cycling tape" refers to making a new magnetic tape file by updating old magnetic tapes.

Current A flow of electric charge which is measured in amperes.

Data sheet Information sheet used to describe a chip's function. Very necessary for the customer.

DCTL Direct-coupled transistor logic. Logic is performed by transistors.

D-type flip-flop A D-type flip-flop propagates whatever information is at its D (data) conditioning input, prior to the clock pulse, to the Q output on the leading edge of a clock pulse. If the data input is 1, the Q output becomes 1 on the leading edge of the next clock pulse. If the data input is 0, the Q output becomes 0 on the next clock.

Debug programs Debug programs help the programmer to find errors in the program while they are running on the microcomputer and to replace or patch instructions into (or out of) the program.

Decimal A system of numerical representation that uses the symbols 0, 1, 2, 3, ..., 9.

Decoder A device used to convert information from a coded form into a more usable form (i.e., binary to decimal, binary to BCD, BCD to decimal, etc.).

Decrement To change the value of a number in the negative direction. If not otherwise stated, a decrement of 1 is usually assumed.

Delay Undesirable delay effects are caused by rise time and fall time, which reduce circuit speed, but intentional delay units may be used to prevent inputs from changing while clock pulses are present. The delay time is normally less than the clock pulse interval.

Demultiplexer Usually, a circuit that converts data from one form to another; a decoder. Usually decodes the input from a specific number of input lines into an output containing more lines. Standard demultiplexers include 4-to-16, 3-to-8, 2-to-4, etc., decoders.

Development system A piece of equipment for the design and test of microprocessor software and the integration of system software and hardware; the development system may be based on the microprocessor under investigation or may merely emulate it.

Device Another name for a transistor, integrated circuit, etc.

Device/gate length Length of the channel of a MOSFET transistor in micrometers. It is the distance between drain and source regions as determined by the 05 mask layer.

Device/gate width Width of the channel of a MOSFET transistor in micrometers. It is the width of the drain and source regions as determined by the 01 mask layer.

Device-size chart A chart often accompanying the logic, describing the size of transistors being used.

Diagnostic programs These programs check the various hardware parts of a system for proper operation, microprocessor diagnostics check the microprocessor, memory diagnostics check the memory, etc.

Die An integrated circuit without the package. Each wafer processed will have many dies.

Dielectric The insulator that makes up a capacitor.

Die per wafer Number of possibly good integrated circuits per wafer.

Diffusion A method of doping or modifying the characteristics of semiconductor material by "baking" wafers of the base semiconductor material in furnaces with a controlled atmosphere of dopant materials. A thermal process by which minute amounts of dopants are deliberately allowed to enter into a substance. Diode junctions, sources, and drains are created by diffusion.

Digit A digit is one character in a number. There are 10 digits in the decimal number system. There are two digits in the binary number system.

Digital A reference to the representation of data by discrete pulses as in the presence of a signal level to indicate the 1s and 0s of binary data. Also, a type of readout in which the data are displayed as discrete, fully formed alphanumeric characters.

Digital circuit Since logic implemented in integrated circuits is binary, a digital circuit implemented in an integrated circuit will be binary logic arranged to perform digital arithmetic.

Direct memory access Direct access to memory locations by a computer.

Discrete Electronic circuits built of separate, finished components, such as resistors, capacitors, transistors, etc.

Donor An impurity that can make a semiconductor N type by donating extra "free" electrons to the conduction band. The free electrons are carriers of negative charge.

Doping The introduction of an impurity into a semiconductor to modify its electrical properties by creating a concentration of N or P carriers.

DTL Diode-transistor logic. Logic is performed by diodes. The transistor acts as an amplifier and the output is inverted.

Dynamic RAM A type of semiconductor memory in which the presence or absence of a capacitive charge represents the state of a binary storage element. The charge must be periodically refreshed.

EAROM Electrically alterable ROM. A read-only memory whose contents may be altered on rare occasion through electrical stimuli.

EBCDIC code A standard code (Extended Binary-Coded Decimal Interchange Code) in which each character is represented by a unique, 8-bit binary code.

ECL Emitter-coupled logic. A form of current-mode logic in which the output is available from an emitter-follower output stage.

EEPROM Electrically erasable PROM.

Emulate To imitate a different computer system by a combination of hardware and software that permits programs written for one computer to be run on another.

Enable A signal condition that permits some specific event to proceed, whenever it is ready to do so. Used synonymously with strobe. A gate is enabled if its input conditions result in a specified output. The specified output varies for different gating functions. In some cases, function enabling inputs allow operation to be executed on a clock pulse after the function enabling input achieves a correct logic level.

Encoder   A device that takes information in one code and encodes it into another (e.g., the decimal-to-binary encoder, decimal-to-BCD encoder, etc.).

Enhancement device   A type of MOSFET that requires a control signal input to turn on the device. The device is "off" when no input signal is present.

Epitaxial layer   A deposited layer of material having the same crystallographic characteristics as the substrate material.

Erase   To remove or clear information, usually used in reference to a computer memory or to equipment such as cathode-ray-tube (CRT) terminals, video screens, etc.

Erasable programmable read-only memory (EPROM)   Similar to a PROM except that the previous information can be erased and new information can be written in.

Etching   The removal of unwanted material, usually by the use of acid.

Exclusive-OR   The output is true only when the two inputs are opposites (complementary) and is false if both inputs are the same.

Exponent of a number   The number of times the base number is to be used as a factor.

Fall time   A measure of the time required for a circuit to change its output from a high level to a low level.

Fan-in   The number of inputs available on a gate.

Fan-out   The number of gates that a given gate can drive. The term is applicable only within a given logic family.

FET   Field-effect transistor.

Field oxide   Protective or insulative layer of silicon dioxide formed on the substrate.

Firmware   Software that is stored in ROM.

Flat pack   An IC package that has leads extending from the package in the same plane as the package, so that leads can be spotwelded to terminals on a substrate or soldered to a printed circuit.

Flip-flop   An electronic circuit having two stable states, and having the ability to change from one state to the other upon the application of a signal in a specified manner. *See the following specific types.*

Flip-flop, D D stands for data. A flip-flop the output of which is a function of the input that appeared one pulse earlier. If a 1 appears at its input, the output one pulse later will be a 1. Sometimes used to produce a one-clock delay.

Flip-flop, JK A flip-flop having two inputs designated J and K. At the application of a clock pulse, a 1 on the J input will set the flip-flop to the 1 or "on" state; a 1 on the K input will reset it to the 0 or "off" state, and 1s simultaneously on both inputs will cause it to change state regardless of what state it had been in. If 0s appear simultaneously on both inputs, the flip-flop state remains unchanged.

Flip-flop, RS A flip-flop having two inputs, designated R and S. The application of a 1 on the S input will set the flip-flop to the 1 or "on" state, and a 1 on the R input will reset it to the 0 or "off" state. It is assumed that 1s will never appear simultaneously at both inputs. (In actual practice the circuit can be designed so that a 0 is required at the S and R inputs.)

Flip-flop, synchronized RS A synchronized RS flip-flop having three inputs, R, S, and clock (strobe, enable, etc.). The R and S inputs produce states as described for the RS flip-flop above. The clock causes the flip-flop to change states.

Flip-flop, T A flip-flop having only one input. A pulse appearing on the input will cause the flip-flop to change states. A series of these flip-flops make up a binary ripple counter.

Fracture Breaking up complex geometries into rectangles rotated on an axis. Used in reference to cell drawing placement.

Gate A circuit having two or more inputs and one output, the output depending on the combination of logic signals at the inputs. There are five gates: AND, exclusive-OR, OR, NAND, and NOR.

Gate, AND All inputs must have 1-state signals to produce a 0-state output (in actual practice, some gates require 0-state inputs to produce a 0-state output).

Gate equivalent The basic unit of measure for digital circuit complexity, based on the number of elementary logic gates that would have to be interconnected to form the same circuit function.

Gate, exclusive-OR gate, NAND The output is true only when the two inputs are opposites (complementary) and is false if both inputs are the same. All inputs must have 1-state signals to produce a 0-state output (in actual practice some gates require 0-state inputs to produce a 1-state output).

Gate, NOR Any one or more inputs having a 1-state signal will yield a 0-state output (in actual practice gates require 0-state inputs to produce a 1-state output).

Gate, OR Any one or more inputs having a 1-state signal will yield a 1-state output (in actual practice some gates require 0-state inputs to produce a 0-state output).

Gate, transistor Control element of an MOS device. Deposited poly-crystalline silicon on top of gate oxide will form a transistor gate. On a drawing, the intersection of the 01 mask with the 05 mask creates a transistor. The 05 mask forms the gate.

Ground, V_{SS} Electrical return path for current from the voltage supply.

Half-add The half-add operation is performed first in doing a two-step binary addition. It adds corresponding bits in two binary numbers, ignoring any carry information.

Handshake A reference to or type of interface procedure that is reasonably easy to implement; its operation is based on a data ready/data received signal scheme that assures orderly data transfer. Can be used as a noun (e.g., "The data exchange was initiated with a handshake."); however, engineering jargon sometimes uses it incorrectly as a verb (e.g., "Data are hand-shaked during the next clock cycle.").

Hard copy Permanent, printed version of information that is other-wise available only on a temporary basis (CRT display data, programs or data ordinarily stored in memory, etc.).

Hardware The electrical, mechanical, and electromechanical equip-ment and parts associated with a computing system, as opposed to its firmware and software.

Hexadecimal notation The expression of a number in a base 16 number system using a combination of digits (0–9) and letters (A–F); the digit/letter combination is a shorthand notation, a compact way of representing a long binary number in 4-bit chunks.

HMOS High-density metal-oxide semiconductor.

Hole The absence of a valance electron in a semiconductor crystal. The movement of a hole is equivalent to the movement of a positive charge.

Hybrid circuit Any combination of two or more of the following in one package: active-substrate integrated circuit, passive-sub-strate integrated circuit, and/or discrete component.

I^2L Integrated injection logic. A bipolar structure characterized by an integrated PNP load device and inverted operation of the NPN logic transistor.

Implantation *See* Ion implantation.

Input A signal that feeds into a circuit.

Input protection A scheme commonly used to ensure that active devices attached to the bonding pads are not damaged by static electric discharges resulting from handling the packaged units.

Instruction cycle The process of fetching an instruction from memory and executing it.

Instruction set The set of general-purpose instructions available with a given microprocessor. In general, different machines have different instruction sets. The number of instructions only partially indicates the quality of an instruction set. Some instructions may be only slightly different from another; others may rarely be used. Instruction sets should be compared using benchmark programs typical of the application, to determine execution times and memory requirements.

Instruction time The time required to fetch an instruction from memory and execute it.

Integrated circuit A collection of circuits all fabricated on a single semiconductor chip. Typical microcomputer ICs include the microprocessor, RAM, ROM, I/O, etc.

Intelligent terminal An input/output device with built-in intelligence in the form of a microprocessor, and able to perform functions that would otherwise require the processing power of the central computer; sometimes called a stand-alone terminal.

Interface The place at which two systems, or a major system and a minor system (such as a computer and a peripheral), meet and interact with each other; the means by which the interaction is effected (e.g., an interface card); also, to connect by means of an interface.

Interpreter A program that executes the instructions from the source (user) language, as each is encountered, and without converting the source language into machine language. An interpreted program is slow—as much as 20 times slower than an assembled program—but speeds up program development because the effect of source changes can be seen immediately.

Inverter A device capable of changing a logic level to its complement, 1 to 0 or 0 to 1.

I/O (input and output) Circuit capable of performing both input and output OUT functions.

Ion implantation Introduction into a semiconductor of selected impurities by means of high-voltage ion bombardment to achieve desired electrical properties in selected regions.

Junction The boundary between a P region and an N region in a semiconductor substrate.

Label Any arrangement of symbols, usually alphanumeric, used to identify an object (e.g., a node, circuit, etc.).

Large-scale integration (LSI) Normally indicates that combinations of complete circuits are integrated on a single chip. Circuit combinations could include counters, registers, gates, timers, etc., all functioning as a single entity, such as found in microprocessors, microcomputers, etc. The actual number of circuits on a chip, to qualify as LSI, is usually in excess of 1000.

Leading edge The leading edge of a pulse is defined as that edge or transition which occurs first (i.e., the leading edge of a high pulse is the low-to-high transition).

Leakage The unwanted flow of current.

LED *See* Light-emitting diode.

Light-emitting diode (LED) A semiconductor diode, the junction of which emits light passing a current in the forward (junction ON) direction.

Line printer A hard-copy device that prints one line of information at a time.

Loader A software program that transfers data and other information from off-line memory to on-line memory.

Location In reference to memory, a storage position or register uniquely specified by an address.

Logic A mathematical approach to the solution of complex situations by the use of symbols to define basic concepts. The three basic logic symbols are AND, OR, and NOT. When used in Boolean algebra, these symbols are somewhat analogous to addition and multiplication.

Logic diagram A symbolic drawing used to express an electronic function.

Long word A word that, due to its length, requires two words to represent it. Long words are normally stored in two adjacent memory locations, totaling 32 bits.

Loop A sequence of instructions that is written only once but executes many times (iterates) until some predefined condition is met. Branch instructions are used to test the loop conditions to determine if the loop should be continued or terminated.

Machine language Instructions and data used directly by a computer without translation.

Majority carrier The mobile charge carrier (hole or electron) that predominates in a semiconductor material.

Mask A glass plate containing an array of photographically produced images corresponding to patterns to be formed on integrated-circuit wafers.

Memory A general term for the computer equipment that holds information in binary language.

Memory address register The microprocessor register that holds the address of the memory location being accessed. *See also* Program counter and Stack pointer.

Memory cycle That portion of the microcomputer timing during which the contents of one location of memory is read out (into a register) and written back into that location.

Memos Short notes written to help one remember something; informal communications.

Microcircuit Another name for integrated circuits.

Microcode A set of control functions performed by the instruction decoding and execution logic of a CPU or microprocessor which defines the instruction repertoire of that computer.

Microcomputer A term used to describe a system using an MPU and, if necessary, additional peripheral units (ROM, RAM, PIA, UART, etc.) to form a complete computer system. Latest microcomputers contain ROM and RAM on the same chip as the MPU.

Microcontroller A term used to describe a small microcomputer (on a single LSI chip) which contains a small amount of built-in memory and I/O devices. Normally used for controlling a minimal number of operating sequences (such as for test equipment), voltage-controlled oscillators (for multichannel radios), etc.

Microelectronics Another name for integrated circuits.

Micrometer Unit of length equal to one-thousandth of a millimeter (the older name was "micron"). Motorola's equivalent to this measurement is 1/25 of a mil and is referred to as a "mocron."

Microprocessor/microprocessing unit (MPU) A device containing the control or arithmetic logic unit of the microprocessor, or the actual microprocessor, including its intimate memory. Normally consists of a single LSI chip, but it could be a group of interactive LSI chips. (Not to be confused with a microcomputer or microcontroller.)

Mil Unit of measurement equal to 1/1000 of an inch.

Mnemonic code Another shorthand, this time a set of symbols that aids the writing of assembly language programs; each assembly language instruction can be represented by a mnemonic.

Mocron Rounded-off micrometer (or micron); 25 mocrons = 1 mil.

Modulus The modulus of a counter describes the number of distinct states which that counter has (e.g., the modulo 10 counter has a modulus of 10 and therefore has 10 distinct states).

MOSFET Metal-oxide-semiconductor field-effect transistor.

Nibble A sequence of 4 bits operated on as a unit.

NMOS Negative-channel metal-oxide semiconductor.

Node The connection of any electrical device. Also used to describe the output side of a device.

NOR gate *See* Gate, NOR.

NOT A Boolean logic term indicating negation. A variable designated "NOT" will be the opposite of its "AND" or "OR" function. A switching function for only one variable. *See also* Inverter.

Octal notation The octal number system is a base 8 system which has eight distinct digits: 0, 1, 2, 3, 4, 5, 6, 7.

Off-line Not being in continuous, direct communication with the computer; done independent of the computer (as in off-line storage).

Ohm Unit of measuring resistance.

Ohms per square Total resistance of a given geometry having four equal sides.

On-line Directly controlled by, or in continuous communication with, the computer (on-line storage); done in real time.

Operand That which is operated on. An operand is usually identified by an address part of an instruction.

Operating system System software controlling the overall operation of a multipurpose computer system, including such tasks as memory allocation, input and output distribution, interrupt processing, and job scheduling.

OR gate *See* Gate, OR.

Output (noun) Information supplied from a gate, data processing system, or computer to other units (terminals, peripherals, RAM, etc.).

Output (verb) The act of supplying (outputting) data from a data processor system or computer.

Package A protective housing that is much stronger and larger than the die, usually made of ceramic or plastic. A package will have leads for electrical connection of the integrated circuit to the outside world.

Passivation A layer of glass placed on the surface of an integrated circuit to protect it.

Photoresist A liquid plastic applied to the wafers that is light sensitive and reacts exactly as film does in a camera. Yellow lighting, however, will not expose the photoresist.

PLA Programmable logic array. A general-purpose logic circuit containing an array of logic gates that can be connected (programmed) to perform various functions.

P/N junction The P/N junction is the interface in semiconductor material of an N-doped region and a P-doped region.

Poly Deposited polycrystalline silicon. Poly is used to form transistor gates, interconnects, capacitors, etc.

Port A signal input (access) or output (egress) point.

Positive logic Logic notation in which the more positive voltage represents the 1-state; the less positive voltage represents the 0-state.

Probe card A special printed-circuit board having a set of leads arranged according to the integrated-circuit pads. The probe card is used to test the integrated circuits while still attached to the wafer.

Probing A term used to describe electrical testing that employs very finely tipped probes applied sequentially to each of the finished dice on a wafer.

Process control language A class of high-level programming language oriented to users in the process industries, and requiring only a minimum of programming skill.

Program A list of instructions that a computer follows to perform a task.

Program counter A microprocessor register that specifies the address of the next instruction to be fetched and executed. Normally, it is incremented automatically each time an instruction is fetched. *See also* Stack pointer and Memory address register.

Programmable gain amplifier (PGA) An instrumentation amplifier that changes its amplification (gain) under command from a digital code supplied through a programmed instruction in software. Very important in allowing a lower-resolution, lower-cost A/D converter to accept a wide-dynamic-range signal.

Programmable read-only memory (PROM) A ROM device that is manufactured with a pattern of either all 1s or all 0s; however, this pattern can be replaced with a specific pattern written into it using a special hardware programmer. Some PROMs, called EPROMs, can have their pattern erased and reprogrammed. PROMs are usually used in the system design stage.

Propagation delay A measure of the time required for a change in logic level to pass from the input to the output of a device, typically measured in nanoseconds (10^{-9} second).

Prototype An original model after which the final model is patterned; a standard or typical example.

Pulldown Refers to a device with the source tied to V_{SS} (ground).

Pullup Refers to a device with the source tied to V_{DD} (power).

Pulse A brief voltage or current surge of measurable duration.

Random access memory (RAM) A static or dynamic memory device that data can be written into or read from; a specific location. The specific RAM location is selected by the address applied via the address bus and control lines. Data are stored in such a manner that each bit of information can be retrieved in the same length of time. This has come to mean, by common usage, read/write memory.

Random logic Common term used in layout describing an area that does not have a particular pattern or is not being repeated.

Ratio Refers to device size-to-device size ratio. *See also* Beta ratio.

Register A device used to store a certain number of bits within the computer circuitry, usually one word. Certain registers may also include provisions for shifting, circulating, or other operations. Registers are normally fast-acting circuits since their speed is directly related to the efficiency of the microprocessor-based system.

Reset *See* Clear.

Resistance The opposition offered by a body or substance to the passage through it of a steady current.

Resolution The smallest detectable increment of measurement. Again, in A/D converters, resolution is usually limited by the number of bits used to quantize the input signal; a 12-bit A/D converter can resolve to 1 part in 4096 ($2^{12} = 4096$).

Restore To return a word to its initial value.

Reticle A 4-inch-square glass plate of individual layers produced by a pattern generator. The image on this plate is then photographically reduced to the proper size to make the actual chip.

Rise time A measure of the time required for a circuit to change its output from a low level to a high level.

ROM Read-only memory. Memory in which the information is stored at the time of manufacture. The information is available at any time, but it cannot be modified during normal system operation. A ROM is programmed according to the user's requirements.

ROM pattern The data stored in a ROM comprise its ROM pattern. The data to be placed in a ROM may be presented as a decimal, hexadecimal, or binary code. It may appear as a pattern of rectangles for an integrated circuit.

Routine Usually refers to a subprogram or task which is less complex than the main program. A program may contain many routines. *See also* Program.

Run To execute a program.

Scale The proportion that an integrated circuit is drawn to the actual size it will be manufactured.

Schematic A transistor-to-transistor wiring diagram. Most integrated circuits are designed from a logic diagram.

Schmitt trigger A fast-acting pulse generator that produces a constant amplitude pulse as long as the input exceeds a threshold value. Used as a pulse shaper, threshold detector, etc.

Schottky TTL A form of TTL logic in which Schottky diodes are used to clamp the transistors out of saturation, effectively eliminating the storage of charge within the transistor, allowing increased switching speeds.

Scientific notation and prefixes

Power	Prefix	Symbol	Common
10^6	mega	M	million
10^3	kilo	k	thousand
10^0	—	—	unit
10^{-3}	milli	m	thousandth
10^{-6}	micro	μ	millionth
10^{-9}	nano	n	billionth
10^{-12}	pico	p	trillionth

Scratch-pad memory RAM or registers that are used to store temporary intermediate results (data) or memory addresses (pointers).

Scribe line The region between integrated circuits on a wafer where the wafer scriber will make its cut.

Semiconductor The word "semiconductor" means acting as an electrical conductor half of the time. Silicon is a semiconductor.

Shift register An element in the digital family that utilizes flip-flops to perform a displacement or movement of a set of digits one or more places to the right or left. If the digits are those of a numerical expression, a shift may be the equivalent of multiplying the number by a power of the base.

Short Common term meaning an electrical connection (usually an unwanted connection, a "short circuit").

Shrink A means of reducing the size at which a chip is manufactured using photographic techniques. A 10% shrink is a linear reduction of each dimension to 90% of the original size.

Silicon A semiconductor. A pure silicon ingot sliced into wafers is the starting material in the production of a chip.

Silicon gate A process using doped deposited polycrystalline silicon to form transistor gates.

Simulation A computer-aided decision process in which proposals are tested before an idea is used on the layout.

Slice Another term for wafer.

SOS Silicon-on-sapphire. A faster MOS technology in which the silicon is grown on a sapphire wafer only where needed. Each device is thus isolated by air or oxide from other devices.

Source language In general, any language that is to be translated into another (target) language; usually, however, it refers to the language used by a programmer to program a system.

Source program Computer program written in a language designed for ease of expression by human beings; symbolic or algebraic. Often applied to assembly language.

SSI Small-scale integration. ICs containing fewer than 10 logic gates.

Stack A sequence of registers and/or memory locations used in LIFO (last-in/first-out) fashion. A stack pointer specifies the last-in entry (or where the next-in entry will go).

Stack pointer The counter, or register, used to address a stack in memory. *See also* Memory address register and Program counter.

Static RAM A type of RAM that does not require periodic refresh cycles, as does dynamic RAM.

Storage *See* Memory.

Strobe A sampling pulse that is used to enable a register, flip-flop, counter, etc.

Subroutine A piece of software that may be used from several locations in a program. A program smaller than the main program, and called up from the main program to perform some specific task.

Substrate The base on which integrated circuits are built. The wafer material forms the substrate material.

Supervisory control An analog system of control in which controller set points can be adjusted remotely, usually by a supervisory computer; also known as a digitally directed analog (DDA) control system.

Supervisory interface In digitally controlled systems, a class of interfaces in which the computer controls the set point of a local controller.

Synchronous Operation of a switching network by a clock pulse generator.

Synchronous operation Use of a common timing source (clock) to time circuit or data-transfer operations. (Contrast with asynchronous operation.)

Thin oxide Insulation material under a polycrystalline silicon gate. Also called the gate region.

Three-state device Any integrated-circuit device in which two of the states are conventional binary (1–0) states; and the third state is a high-impedance state. When in the third state, these devices present a high impedance to their respective output lines to reduce power drain and allow access to the common bus lines by other devices.

Threshold The input voltage at which the output logic level changes state.

Threshold voltage Theoretical turn-on voltage of a transistor.

Topology A branch of mathematics that deals with the properties of a geometric figure that does not vary when the figure is transformed in certain ways. In MOS technology, the term "topology" refers to the layout of a transistor, and the layout of an entire MOS circuit.

Trailing edge The trailing edge of a pulse is that edge or transition which occurs last. The trailing edge of a HI clock pulse is the HI to LO transition.

Transistor The basic solid-state device used to amplify or switch electrical current.

Truth table A tabular chart that lists all the possible combinations of the inputs and outputs of a circuit.

TTL Transistor-transistor logic. A modification of DTL which replaces the diode cluster with a multiple-emitter transistor.

Tunnel N regions used to conduct signals under a metal line. The region is insulated from the metal by an oxide layer.

Universal asynchronous receiver-transmitter (UART) A device that provides the data formatting and control to interface serial asynchronous data. Input serial data are converted to parallel data for transfer to the microprocessor or microcomputer via the data bus. Conversely, output parallel data, from the data bus, is converted to serial data in the output. (Similar to Motorola's Asynchronous Communications Interface Adapter.)

V_{DD}**—power** Dc voltage normally applied to the drains of load devices.

VLSI Very-large-scale integration. VLSI devices are ICs that contain 10,000 or more gate equivalents.

Volatility With respect to memory, an inability to retain stored data in the absence of external power.

Volt Unit of electromotive force.

Voltage Electric potential or potential difference expressed in volts.

V_{SS}—**ground** Electrical return path for current from the supply voltage.

Wafer The thin slice or flat disk of pure silicon cut from a single-crystal ingot on which integrated circuits and other semiconductor devices are fabricated.

Word The term "word" denotes an assemblage of bits considered as an entity for manipulation (read in, stored, added, read out, etc.) by the microcomputer in a single step. Common usage has defined a word as 16 bits. Two types of words are used in every microcomputer: data words and instruction words. Data words contain the information to be manipulated. Instruction words use the microcomputer to execute a particular operation.

Write To put information into a storage device (RAM, register, etc.). This can be either a verb (to signify the writing action) or a modifier (to signify the write function, write signal, etc.) or a noun (as in the case "strobe is generated by a write to the input data register.").

Yield Final number of good chips per wafer.

INDEX